Engineers as Executives

IEEE PRESS
445 Hoes Lane, PO Box 1331
Piscataway, NJ 08855-1331

IEEE PRESS Editorial Board
William Perkins, *Editor in Chief*

R. S. Blicq	J. D. Irwin	J. M. F. Moura
M. Eden	S. Kartalopoulos	I. Peden
E. M. Etter	P. Laplante	E. Sánchez-Sinencio
G. F. Hoffnagle	A. J. Laub	L. Shaw
R. F. Hoyt	M. Lightner	D. J. Wells

Dudley R. Kay, *Director of Book Publishing*
Carrie Briggs, *Administrative Assistant*
Lisa S. Mizrahi, *Review and Publicity Coordinator*

Deborah J. Graffox, *Production Editor*

Cover Design by
Tisa Lynn Lerner / Lerner Communications

Also of Interest from IEEE Press

SPARKS OF GENIUS: Portraits of Electrical Engineering Excellence
By Frederik Nebeker, *IEEE Center for the History of Electrical Engineering*

This biography chronicles the ground-breaking technological advancements made by eight of this century's most brilliant engineers.

1994   Hardcover   280 pp   IEEE Order No. PC0382-2   ISBN 0-7803-1033-0

TECHNOLOGICAL COMPETITIVENESS: Contemporary and Historical Perspectives on the Electrical, Electronics, and Computer Industries
Edited by William Aspray, *IEEE Center for the History of Electrical Engineering*
1993   Hardcover   384pp   IEEE Order No. PC0324-4   ISBN 0-7803-0427-6

THE LEADER-MANAGER: Guidelines for Action
By William D. Hitt, Ph.D.
1993   Softcover   280pp   IEEE Order No. PP0356-6   ISBN 0-7803-1007-1

MANAGEMENT IN ACTION: Guidelines for *New* Managers
By William D. Hitt, Ph.D.
1993   Softcover   304pp   IEEE Order No. PP0357-4   ISBN 0-7803-1008-4

Set of Both Volumes
1993   Softcover   584pp   IEEE Order No. PP0361-6   ISBN 0-7803-1012-8

Engineers as Executives

An International Perspective

William Aspray
IEEE Center for the History of Electrical Engineering

with the assistance of
Jill Cooper

IEEE PRESS

The Institute of Electrical and Electronics Engineers, Inc., New York

This book may be purchased at a discount from the publisher when ordered in bulk quantities. For more information, contact:

IEEE PRESS Marketing
Attn: Special Sales
P.O. Box 1331
445 Hoes Lane
Piscataway, NJ 08855-1331
Fax: (908) 981-8062

© 1995 by the Institute of Electrical and Electronics Engineers, Inc.
345 East 47th Street, New York, NY 10017-2394

All rights reserved. No part of this book may be reproduced in any form, nor may it be stored in a retrieval system or transmitted in any form, without written permission from the publisher.

The brand and product names referred to throughout this book are the trademarks and registered trademarks of their respective companies.

Printed in the United States of America

10 9 8 7 6 5 4 3 2 1

ISBN 0-7803-1103-5
IEEE Order Number: PC 04564

Library of Congress Cataloging-in-Publication Data

Aspray, William
   Engineers as executives : an international perspective / William Aspray ; with the assistance of Jill Cooper.
    p.   cm.
   ISBN 0-7803-1103-5
   1. Engineers. 2. Executives. 3. Industrial management.
I. Title
TA157.A799 1995
620'.0068—dc20
                                            94-21224
                                              CIP

Contents

	Introduction	vii
Part I	**Germany**	**1**
	Chapter 1–Herbert Bruch	
	About Grundig AG	3
	Interview	5
	Chapter 2–Ernst Denert	
	About sd&m	17
	Interview	19
	Chapter 3–Kurt Schips	
	About Bosch GmbH	59
	Interview	61
	Chapter 4–Arno Treptow	
	About AEG	77
	Interview	79
Part II	**Japan**	**91**
	Chapter 5–Katsutaro Kataoka	
	About Alps	93
	Interview	95
	Chapter 6–Koji Kobayashi	
	About NEC	111
	Interview	113

Chapter 7–Kazuhiko Nishi
About Ascii 125
Interview 127

Chapter 8–Takashi Sugiyama, Takashi Yamanaka
About Yokogawa 149
Takashi Sugiyama—Written Answers 151
Takashi Sugiyama—Interview 157
Takashi Yamanaka—Interview 177

Part III United States 195

Chapter 9–Robert Galvin
About Motorola 197
Interview 199

Chapter 10–Mitchell Kapor
About Lotus 219
Interview 221

Chapter 11–Arthur P. Stern
About Magnavox 243
Interview 245

Chapter 12–Erwin Tomash
About Dataproducts 269
Interview 271

Introduction

Managers of technological businesses face great challenges. They must not only be able to make difficult decisions faced by any business leader operating in a highly competitive international market, but also handle technologies that are often arcane, expensive, and rapidly changing. What qualifications prepare someone for this kind of executive position, and what kinds of challenges are faced every day? What impact does an engineering background have on their management philosophy and practice? Are these experiences universal, or do they vary from continent to continent, and from country to country?

To gain insight into these questions, the IEEE-Rutgers Center for the History of Electrical Engineering conducted oral histories with thirteen senior executives from the computer, electrical, and electronics industries. Through an examination of their personal histories, we investigated general issues about the challenges and styles of managing technological businesses.

We attempted to obtain as wide a cross section of views as possible within our limited project. We chose one country to represent each of the world regions most active in technological business today: Japan representing Asia, Germany representing Europe, and the United States representing the Americas. All of the executives we interviewed held senior positions in successful technological businesses, but in many respects their companies varied widely. Companies ranged in size from small entrepreneurial start-ups to large corporations employing more than one hundred thousand employees. Their products ranged from software to components to finished systems. Some worked in commercial markets; others primarily in military markets.

A standard set of questions was prepared in advance of the interviews (with slight modifications to the question set for executives in the software business). Questions pertained to very general (not company-specific) issues:

- the importance of a technical background in managing the company
- qualifications needed to be a top manager
- procedures and methodologies employed within the company
- importance of quality, reliability, and maintainability issues to the company's well-being
- knowledge needed about customers' needs and operations in order to run their business successfully
- the role of service in their business
- methods for achieving economies of scale
- relations to other industries
- role of the government in their business
- recruitment and training of managers
- importance and means of continuing education
- how to deal with the rapid pace of innovation
- general management lessons

The questions were shared with most of the executives prior to the taped interview. In some cases the interview focused primarily on answers to these questions, but in other cases the overall structure for the interview was based on the person's career. In every case, however, a number of these issues was raised and at least two or three were discussed at length. (In one case, Dr. Sugiyama wrote answers to these questions in advance of the interview, so the interview amplified on his answers and focused on other issues. His written answers precede his interview transcript.)

The transcripts that appear in this volume have been edited by both the Center's staff and the executives themselves. As an aid to the reader, each transcript is preceded by a brief history of the principal company under discussion in the interview. Most of the interviews also include a portrait photograph of the executive.

The first section of the book includes interviews with four German executives, beginning with Herbert Bruch, a member of the Management Committee that runs the Grundig Company. Grundig AG is a worldwide producer of consumer electronics, such as televisions, compact disc players, video recorders, upmarket hifi components, world band radios, car radios, and telephones, as well as industrial electronics, office electronics, and information technologies. Bruch discusses the way in which research and development, manufacturing engineering, and product development are organized within Grundig. Other topics include the strong position of European companies in logistics and customer service, the value of joint ventures, the relations between Grundig and its parent company, Philips, the importance of continuing education, and the difficulty of training engineers to be managers. He gives special emphasis to the importance of maintaining engineering capability within Grundig and the need for senior operations managers to keep track of what is happening on the factory floor.

The second interview is with Ernst Denert, managing director and one of the co-founders of sd&m, a leading custom software design and consulting firm with a distinguished client list that includes Lufthansa, AEG, Siemens, Thyssen, TUI, BMW, and the Deutsches Bundesbahn (national railway). One of the most interesting topics discussed in this interview is how Denert's participation in the Berlin student protests of 1968 shaped his career and the culture of the company he founded. Denert also discusses how his philosophy of software design and management have shaped his company's operating practices. He considers the "software crisis," the tools and methods developed to overcome it, and why universities have difficulty providing adequate training in software engineering. Other topics include the lack of risk in custom software entrepreneurship experienced by Denert, the ethical responsibilities of spinoff companies to their parent organizations, the strengths and limitations of being a small company, and strategic reasons for opening a small business to new partners.

The next interview is with Kurt Schips, a member of the Management Committee that runs Robert Bosch GmbH, an international company with more than 180,000 employees that has been a leader in the automotive components industry since the nineteenth century and today produces mechanical automotive equipment, communications technology, power tools, household appliances, thermal technology, plastic products, packaging machinery, industrial equipment, and hydraulic and pneumatic machinery. Schips discusses competition among Japan, Germany, and the United States and some of the factors that affect the outcome, such as the tension between short-term return on investment and market share, and the effect on research and development of the comparatively longer Japanese work year. He considers at length the organization of research and development, manufacturing engineering, and product development. Other topics include decentralization, the importance of building new market areas on existing expertise, the value in early market entry, training engineers to be managers, and the importance of technical training for senior managers.

The final German interview is with Arno Treptow, a member of the Management Committee of AEG. AEG has more than a century of experience in the electric power business and today, as one of the Daimler Benz companies, has international businesses in power transmission, automation, railroad systems and components, domestic appliances, and microelectronics. The first part of the interview discusses the structure of the international market in electric power, the shakeout in that industry over the past forty years, and the particular importance of customer-supplier relationships in this market. Treptow considers the changes that occurred in management philosophy when Daimler acquired AEG and the ways in which various functions are centralized or vested in the divisions today. Other topics include the effect of the breakdown of the Iron Curtain on European business, the training programs at AEG for teaching engineers to manage, and the backgrounds found among senior managers.

The first interview in the second section is with Katsutaro Kataoka, a highly individualistic corporate leader who, as a World War II veteran, built up a small firm manufacturing variable capacitors into Alps Electric, one of the world's leading, horizontally-diversified secondary components manufacturers. Alps built many of the television tuners that enabled the Japanese to dominate the television manufacturing industry, developed a major car stereo business (Alpine) out of a supply relationship with Honda Motors, and today supplies floppy disk drives for Apple, IBM, and other companies. Much of the interview concerns how Kataoka was able to build his small postwar venture into a major, independent supplier of electronic components, especially in the face of the stronghold that the large Japanese systems manufacturers have over their suppliers. Of particular interest is Kataoka's explanation of how he learned his operating principles and his commitment to quality manufacturing in the 1950s and 1960s from American companies, which themselves later abandoned these principles. Other topics include the value in situating factories in rural areas, managing rapid growth, the move from customized to standardized products in order to remain competitive, the Vietnam War as a watershed in Japanese-American business relations, and the future importance of China as market and producer.

The next interview is with Koji Kobayashi, Chairman Emeritus of NEC Corporation, which during his tenure as company leader became the world's largest producer of computing and telecommunication technologies. NEC's activities were shaped by Dr. Kobayashi's vision of the integration of computer and communications technologies into what he has called C&C. In the interview Dr. Kobayashi discusses the importance of this integrated approach to technology, the problems of building up a business after the Second World War, the difference between a technical and a financial mindset in the leadership of technological businesses, and how to operate a multinational company in countries other than that of the home office.

The third Japanese interview is with Kazuhiko Nishi, the inventor of the first laptop computer and the joystick used in millions of Nintendo video games. His company, the Ascii Corporation, today has a well-integrated and future-oriented group of business activities in personal computing, entertainment, business workstations, semiconductor design, and on-line services. In this wide-ranging interview, Nishi discusses many aspects of his business philosophy: communication between senior and middle level managers, decentralization and optimal size of profit centers, importance of research and development to long-term growth, joint ventures as a way to overcome lack of mobility of Japanese engineers, integrating business areas intelligently for long-term growth, rating an engineer's skills, recruitment of software designers, the importance of constant product enhancement when product cycles are short, difficulties of fast-tracking employees in Japan, and the importance of dominating one's market niche.

The final two Japanese interviews are with executives from the Yokogawa family of companies. One is with Takashi Sugiyama, now the chairman of the board of Yokogawa Research Institute Corporation. He was the first president of Yokogawa Medical Systems, a joint venture between Yokogawa Electric and General Electric to manufacture medical diagnostic systems such as computer-aided tomography, nuclear medical imaging, and ultrasound systems. He discusses how the company was formed as a marketing arm for General Electric, but through strong production engineering and an emphasis on service became a leading worldwide producer in this market area. Other topics include organization of R&D, the research planning process, just-in-time and total quality control manufacturing systems, academic-industry alliances, and continuing education.

The other interview is with Takashi Yamanaka, the president of Yokogawa Electric Corporation, the parent company, which is a leading international supplier of measurement and control instrumentation and information management systems. Mr. Yamanaka discusses how to achieve successful joint ventures, such as Yokogawa Electric has attained with Hewlett-Packard and General Electric. He also considers strategies for managing overseas subsidiaries, the growing market in China, Japanese-American business relations, American strength in innovation versus Japanese strength in production engineering, and the "boomerang effect" that occurs when developing nations begin to export technology they had previously imported.

The third section comprises four interviews with American executives. The first is with Robert Galvin, the chairman of the Executive Committee at Motorola, one of the leading worldwide manufacturers of semiconductors and a major supplier of information systems, communications technologies, and automotive, industrial, and defense electronics. Galvin discusses such topics as the importance of integrated management and engineering, the role of quality as a principal organizing force within a technological business, the importance of the Malcolm Baldrige Award, and international competition, especially with the Japanese.

Mitchell Kapor, one of the founders of Lotus Corporation, is the second American executive interviewed. Lotus is one of the largest and fastest growing software companies in the world, known for its integrated products for microcomputers. Kapor's main topic of discussion is the problems of managing hypergrowth when a small start-up firm is wildly successful. In this context he discusses the particular difficulties with scaling up marketing and development, the kinds of technical, people, and communication skills desirable in the chief executive officer, and the problems of corporate culture clash that can occur when people with different backgrounds are thrust together so rapidly. Other topics include the differences of operating in an emerging versus a mature industry, how to design a successful product, the importance of timing in market entry and product delivery, the differences in managing small-scale and large-scale organizations, and the classic mistakes made by entrepreneurs when they start their second company.

The next interview is with Arthur Stern, who was president of Magnavox Advanced Products and Systems Division and who held senior positions prior to that in General Electric, Martin-Marietta, and Bunker-Ramo Corporation. Magnavox, now owned by North American Philips Corporation, is a major consumer electronics manufacturer, the world's leading supplier of handheld radios, and a major producer of military and commercial products in communications, antisubmarine warfare, infrared systems, satellite navigation, and tactical information systems. Stern discusses the relationship of the parent company to its subsidiaries, the need for and problems in technological forecasting, assessing customer needs to improve product acceptance, customer lock-in and bidding practices in the defense electronics industry, and the difficulties of converting defense operations to commercial activities.

The final interview is with Erwin Tomash, founder of Dataproducts Corporation, one of the world's leading manufacturers of computer printers. Prior to founding Dataproducts, he held senior management positions at Telemeter Magnetics and Ampex. The interview considers issues confronted by small engineering firms in new technology areas, such as transforming a project team into a business with marketing and financial capabilities; establishing methods to achieve performance, uniformity, and reliability in new products; moving from innovative design to volume manufacturing; the problems in obtaining effective middle managers to run projects; and the need for senior managers to have good technical training until the technical area matures. Other topics include establishing budgets and product pricing, relations between suppliers and systems purveyors, tension between capital investment and short-term return in publicly held companies, differences between manufacturing engineering and development engineering approaches to product development, shortcomings of the venture capital system in America, and the role of the government in American high technology.

Many people have generously assisted with this project. When I became ill unexpectedly, my colleague Andrew Goldstein kindly stepped in to conduct the interview with Mitchell Kapor. James Gover, an IEEE Congressional Fellow, helped to formulate the basic question sets. My research assistant, Jill Cooper, prepared the company histories that precede each interview and edited most of the transcripts. Oskar Blumtritt of the Deutsches Museum spent many hours identifying executives and arranging my interviews in Germany; and Yuzo Takahashi of Tokyo University of Agriculture and Technology played a similar role in Japan. Eiju Matsumoto of Yokogawa Electric, Kiyoshi Yamauchi and Michiyuki Uenohara of NEC, Noriko Kase of Ascii, and Hirotoshi Okamura of Alps Electric all provided valuable assistance within their companies. Thanks also to Liz Roach and Colleen O'Neill for their outstanding job with transcription and to Michael Ann Ellis for her work on editing and administration.

I am most grateful to these thirteen executives for making time in their busy schedules to be interviewed. Finally, let me thank the IEEE Foundation and the AT&T Foundation, which generously provided the funding for this project.

Part I
Germany

Chapter 1

Herbert Bruch

About Grundig AG

Max Grundig, who was born in 1908 in Nürnberg, went into business for himself at an early age with a radio sales and repair company (*Radio Vertrieb Fürth*) that began business in 1931. Despite the difficult economic times, the company prospered. When the war came, the company began making transformers and later electric detonators, and it grew at a fantastic pace, employing six hundred people by war's end.

After the war, Grundig obtained a license from the U.S. occupying forces to manufacture transformers and measuring instruments, but soon found a large market in radios. By 1949 Grundig's company was the largest German manufacturer of radios, and it soon expanded its product line, making tape recorders (the first in the world for home use), phonographs, and later television sets.

The company changed its name to Grundig in 1948 and later became a common-stock company. The number of employees grew steadily in the postwar decades to thirty-five thousand in the late 1970s. In 1960 the first factory outside of Germany was established in Northern Ireland. There followed others in Italy, France, Portugal, Austria, Taiwan, and Spain. In addition, licensed production facilities were established in several other countries.

Grundig AG has utilized its strong commitment to research and development in the electronics industry to become West Germany's largest consumer electronics firm. Its popular consumer electronics products include: television, video recorders, camera recorders, hi-fi equipment, audio equipment, car radios, and telephones. In addition to consumer electronics, Grundig has an industry electronics division, an office electronics division, and a division for PC-data technology.

In the early 1980s Philips acquired a 24.5 percent stake in Grundig and continues to be Grundig's parent company today. Grundig currently employs 20,473 at home and in subsidiaries across Europe, Malaysia, and Hong Kong. In 1992 the company's annual sales exceeded DM 4.2 billion.

Herbert Bruch

Place: Fürth, Germany

Date: July 2, 1993

Bruch: My name is Herbert Bruch. I am a member of the board of Grundig AG in Germany. I got my Ph.D. in physics in 1964.

Aspray: Where did you train?

Bruch: I studied at different universities. My background is ultrasonics and acoustics. This was the reason why after some steps I started with basic R&D in Grundig. During my company career I have held several functions on both technical and management levels.

Aspray: Would you take five minutes to go through the path of your career for me please?

Bruch: I started in 1964 in the basic R&D laboratory working on electroacoustics, remote controls, ultrasonic microphones, etc. I developed ultrasonic delay lines for TV applications. The first field effect transistors appeared and I started developing memories. This work was successful. For this reason I got the opportunity to transfer the R&D results into mass production.

After this I was involved with electro-optic products such as traffic detectors, picture transmissions through light beams and kerr cell modulation. These experiences enabled me to be the head of an industrial engineering department with a tool and mold shop, electrical and mechanical equipment for production, and a chemical laboratory. Later I became plant manager for a PC board factory, and was given responsibility first for the video recorder production and later for video and TV production. At the age of thirty-nine I became a member of the supervisory council of Grundig AG. At forty-two I became a member of the board of management.

Aspray: So young!

Bruch: At that time in European companies this was really young. As a member of the management board I was responsible for production of audio, TV, and video recorder products. The suggestion was made to combine production and development. So I became responsible for development and production of TV and audio products, such as hi-fi, SE-receivers, and car audio. After a company reorganization into business groups I became responsible for the complete TV business, including development, preproduction, components, R&D and production, and the commercial activities. In my present job as a board member, I am responsible for R&D and production of TV sets, satellite receivers, satellite technology and preproducts.

Several months ago the corporate purchasing department was also integrated. Sixty to eighty percent of the value of our products are purchased items. So this makes sense. This is my actual job. I am now fifty-four. I am a member of several supervisory councils in sister companies and some social organizations closely related to our company, such as the German *Berufsgenossenschaft für Feinmechanik und Elektrotechnik,* an organization which covers social and safety aspects in industry.

Aspray: Let's talk first about R&D. Can you tell me how R&D is organized in the company? Is it centralized?

Bruch: R&D and development are separate functions. R&D is a central task within the company. Product development is organized in the different business groups.

Aspray: Is the basic R&D that is centralized paid for out of a centralized fund? Or is it paid for by the division that requires the research?

Bruch: Actually, we have a central budget for R&D. However, the costs have to be covered by the business units, who are the owners of the business and who have the best and most effective orientation to the customer.

Aspray: To what degree do those people in the business units set the research agenda for the basic research group?

Bruch: The R&D group is not very large in our company because we have no R&D for ICs. We have no R&D for picture tubes. We buy these items. For this reason, our R&D staff receives fifty percent of their orders for future products from the business groups and the remaining fifty percent comes from the BOM. Typical products for R&D are digital TV, HDTV, RDS, DSR, image sensor displays, and product ideas from the professional electronic sector.

Aspray: What kinds of work do they do?

Bruch: They develop prototypes of numerical control for tooling machines, all kinds of supervision technologies in subways, banks, etc. They are also involved in the development of test equipment for laboratory uses and test equipment for the car industry, such as gas analyzers. They work on new business and new activities in the environmental field, such as recycling or the control of water in cleaning processes. In their work they use knowledge of TV and VCR technologies. They develop data storage using technology developed in our VCR department. If the preproduct looks successful, it goes into product development and later into production.

Aspray: Suppose the centralized R&D group comes up with a promising idea that might have some practical value to the company. What happens next? What is the procedure for moving it towards the product?

Bruch: First we do some brainstorming on the project. Then the project team works out a business plan together with the marketing and sales people. This plan is presented to the business group for a decision.

Aspray: So you do some market research? Is that right?

Bruch: Naturally. Market research is part of the business plan and we use both internal and external marketing people.

Aspray: Do the people who come up with an idea move with it from the central R&D group?

Bruch: Not generally. Naturally, these people have to keep in touch with the problem and with the project. But usually with a prototype or specification the project moves into the product division. Then the idea becomes a development project which is budgeted and monitored. There is a master plan and a plan for market introduction.

Aspray: Is manufacturing engineering done in the development group?

Bruch: Actually no. However, in the TV business we are currently moving manufacturing engineering into the development group. Normally, development and production have to be kept as close together as possible. Since the product families are produced in different locations, it makes sense to keep product development and manufacturing engineering close to the leading factory. This factory provides assistance to the satellite factories.

Aspray: What is the typical life of one of your products?

Bruch: In the fast-moving consumer industry, the classical product lifetime is one to two years.

Aspray: So you must have several replacements in the works at any one time for every given product.

Bruch: There are products with longer cycle times, such as satellite receivers. In semiprofessional and professional applications, cycle times are even longer. But every product undergoes a modification during its lifetime. The cycle follows a bell curve. The cycle time depends on the product. The features may remain untouched for a longer period but optical redesigns are made and the cost-price must decrease. A family of TV sets has a lifetime of approximately two years in the market including optical and cost modification.

Aspray: That doesn't give you enough time to make incremental improvements in your manufacturing technology for a product, does it?

Bruch: Yes. I would say the current learning curve for our products is about half a year. Moreover, a TV set is followed by a new TV set, so the basic technology is the same. It is only the big changes, such as the change from analog to digital technology, or from wooden cabinets to plastic cabinets, where the products change substantially.

Aspray: How does the field information come back into the loop for product development?

Bruch: We have a marketing and sales organization. These people provide input. We have additional sources looking for product ideas. The marketing organization has its ear close to the market, close to the customer. The second source is the IC, the display, and the component industry. This input is covered by the R&D people. Our own research results are also used as a basis for new products. The product description results from the optimal combination of all sources of input. This product description is analyzed and after the evaluation by the business groups the product is prepared for the market. After market introduction we monitor the product and thus we are well informed as to whether the idea is going to be successful or not.

Aspray: In what ways do you think the approach of your competitors is different?

Bruch: We know what our competitors are doing. We know what our customers want. We are fast in making decisions and we have clever engineers. As a European company we are strong in logistics and customer service. Our production is flexible, highly automated, and equipped with advanced technology. We have subcontractors in Hungary to compensate for the cost of production in a high-cost country like Germany.

Aspray: It is an opportunity in the sense that you have new markets and less expensive labor available to you. But it is also a challenge because you have expensive labor here in what was West Germany.

Bruch: Yes. This is exactly right.

Aspray: Are you actually closing plants in what was West Germany and moving operations to less expensive places?

Bruch: Yes, we have to do that. There is overcapacity in this industry and we have to reduce it.

Aspray: You mentioned joint ventures just a moment ago. When I was reading on the history of the company, I noticed that it has had a number of joint ventures over time. What were Grundig's reasons for entering into joint ventures?

Bruch: To combine technological, economical, and market possibilities.

Aspray: Could you elaborate on that point, please?

Bruch: If a company is the world leader in a special technology and you need that technology, then it makes sense to join them. Capital investment in new activities is expensive. Cost sharing with a partner is a good alternative. If economy of scale is a criterion for better product prices, it makes sense to join forces and to produce together. In a joint venture you maintain ownership of the technology and you are not too dependent on other companies.

Aspray: One reason you didn't give was an inability to distribute or market in a particular geographical region. Is that a problem?

Bruch: Yes.

Aspray: Is that an actual reason you enter into joint ventures?

Bruch: This is more on the commercial than on the technical side. We have business partnerships in Indonesia, Asia, South America....

Aspray: Rather than entering into joint ventures, do you cross license a fair amount?

Bruch: Yes. We have cross licenses with our mother company and with other companies. But this is mainly related to patents.

Aspray: You said earlier that a very substantial portion of the cost of your consumer products is in the components that go into them. What is your relationship with your suppliers of these components? When you are designing a new product do you get involved very closely with them? Just how does that work?

Bruch: We have partnerships for basic components and preproducts. These are close technical cooperations. Exchange of information and specifications with our main suppliers in Europe and in Asia forms the base of our business.

Aspray: Does that mean that over time you have developed special working relationships with a single or a couple of companies, say IC companies?

Bruch: With a couple. We have historically good contacts with our mother company, Philips, but we also have excellent relations with other leading companies in our industry all over the world.

Aspray: When you are looking at display units or ICs, do these have to be specially made for your products? Or are these components that are more or less just available?

Bruch: We use both standard and customized components.

Aspray: In the case of customized components, do the people from your R&D team, or product development team, invite members of the display team from the supplier to join in the design discussion at an early stage?

Bruch: Yes, and this is very important for success.

Aspray: Is this a kind of mutual design effort?

Bruch: This is the intention. R&D is expensive. To reduce costs certain agreements have to be made. For this reason a long-term partnership is needed. This kind of cooperation is necessary and has to be negotiated at an early stage of product development.

Aspray: In a business such as the television business, where you have high numbers of products coming off of your lines, are manufacturing technologies more important than with customized big expense products?

Bruch: This is absolutely right. The product definition itself, of our side, is designed electrically, mechanically, and optically under license.

Aspray: But, in a way, you are still a value-added manufacturer. Even if these components have a lot in them already, you still have to put them together. You are putting a lot of them out the door every day. So you still must have some concerns about your manufacturing. Can you tell me about the kinds of issues that arise in improving manufacturing?

Bruch: Improvements to manufacturing can only be made integrally. Improvement hinges on development of excellent concepts. Improvement rests on components and material and labor content. In the industrial process we demand maximum efficiency every year. A very important part of our industrial process is logistics.

Aspray: Is that something that has become of greater importance to the company over time?

Bruch: Yes. Every couple of years we have to recheck our policy to ascertain the appropriate rate of integration. We have to adapt technology and the degree of vertical integration. I have already mentioned the change from analog to digital technology, which has consequences in testing, aligning, and handling. One aspect which I did not emphasize enough before is the role of logistics in the production process, which is very important for the time needed to market a product. Flexibility in our final products is an additional aspect. The customers want the product immediately after their order is placed. Product design and industrial feasibilities have to follow these demands.

Aspray: Since you are maintaining within the company the engineering aspects of your products, does that mean that you need specialized test equipment or specialized design tools? Do you have to build them yourself or are they available?

Bruch: We manufacture special test equipment for our products ourselves. Standard test equipment is purchased externally. Mechanical production line equipment is designed in-house and then purchased. We adapt standard machines to our special applications. We manufacture 30 to 40 percent of the tools and molds ourselves.

Aspray: Can you give me an example of the kind of thing that you would either have to adapt or build for yourself?

Bruch: For example, a plastic molding injection machine is a standard product. The application of robots, the design of the product, tool making, multimachine operation, and process control are modified, optimized, and adapted to our requirements.

Aspray: Can you tell me about your relations with Philips? Do you work as a profit center? What kinds of research are shared? What kind of management is shared? Which functions are shared? How does this operate?

Bruch: It is a partnership. With 32 percent share holding in Grundig and a management contract, Philips is responsible for Grundig. This management contract gives Philips all necessary responsibilities. Since 1993 the Grundig annual profit and loss account is consolidated with that of the Philips Company.

Aspray: Do you share any personnel or costs? Do you do any marketing jointly with them?

Bruch: We can use Philips resources. We have a cross-license agreement. We use Philips components. We cooperate on the industrial activities but in the marketplace we are competitors.

Aspray: Tell me about your personal day-to-day duties. What are the kinds of issues that you confront on a daily basis?

Bruch: Fifty percent is strategical and planning work. Twenty percent is purchasing. The rest is day-to-day business.

Aspray: Which duties are the most challenging? What I am going to eventually try to get at is what role your technical background plays in the decisions that you make in your job.

Bruch: Managers often state that they spend most of their time working on strategic visions. In many cases the truth is a little bit different. I manage our work processes and methods with the objective of structuring them in such a way as to achieve optimum efficiency while at the same time meeting customer demands and being profitable. But I also have to do some day-to-day business.

I have not lost contact with the floor. I think that this is important especially on the technical side. My commercial colleagues must not lose contact with the market. I have to keep an eye on what is going on in our industrial business. This includes concept, production place, vertical integration, material bill and labor cost, quality aspects, and customer service. I have to travel a lot. I have to keep in touch with our partners, competitors, subcontractors, and our suppliers, and I also have to maintain close contact with the line managers in the factories.

I see the managers and the people on the floor. I think it is a rather good mixture. Fifty percent strategic and 50 percent operative. That is what I do personally.

Aspray: To what degree does your technical background enable you to do your job better?

Bruch: I think that a technical background is a must in my job. My R&D background has taught me to think and to analyze. This enables me to check quickly whether our plans are correct and successful and will lead to successful products. My progress through the whole organization in the course of my career has provided me with a knowledge and a feeling for the important things in the company.

Unfortunately I have never worked in another organization for any length of time. My basic experience is in-house. I have traveled frequently to the Far East, the U.S., and other countries where electronic things happen. I have taken part in various seminars and training courses to improve my knowledge. Overall, if I could give a piece of advice to a young scientist at the bottom of the career ladder, I would recommend he obtain experience in at least two or three companies. In my particular case the company provided me with all opportunities and a big enough playground.

Aspray: Do the other members of the board have technical backgrounds as well?

Bruch: On the Grundig BOM there are three colleagues with a technical background and four with a commercial background.

Aspray: The commercial people, are they in jobs of a different character?

Bruch: They occupy posts in finance, personnel, marketing, and sales. There are also commercial people in charge of integral business.

Aspray: How do they deal with these problems of understanding and making decisions about the technology?

Bruch: They are, for example, chairmen of important business groups and they are supported by technical staff within their organization.

Aspray: What about the chief executive officer of the company?

Bruch: The CEO of our company has a commercial and legal background. He is a forward-thinking man with a completely business orientation but he also has good technical knowledge and a feeling for our products.

Aspray: Do you think this is a successful combination?

Bruch: Yes, this works perfectly. Most European companies which are not extremely technology driven are managed by commercial people. This makes sense.

Aspray: When the company is hiring young engineers, do you have trouble getting qualified people?

Bruch: Not at the moment. We have high unemployment in Europe. Even good, qualified engineers have problems finding positions now. We hope that this is a temporary situation.

Aspray: Is the university education that these people have appropriate for the job that you ask them to do?

Bruch: On the technical side, our R&D department is mainly staffed with people who have earned a university degree. Our engineers with a polytechnic background tend to work in product development and industrial engineering. We also have some technicians who we trained in-house.

Aspray: With continuing education programs inside the company?

Bruch: On an internal and external basis.

Aspray: These are things that are continuing year after year?

Bruch: Training programs are conducted year after year if there is a demand.

Aspray: For a typical employee, how many days a year would they spend in continuing education?

Bruch: It depends on the level. It is difficult to give an average figure. I, myself, spend approximately ten days per year on education, participating in seminars and training programs. But all of our staff continuously undergo internal training.

Aspray: How do you train engineers to be managers within the company?

Bruch: It is very difficult. We use both internal and external programs to train them. We hire external teachers to transfer their experience to our staff and we also make use of the Grundig Academy. For example, we have a program called Train the Trainer. This is an activity of which we are particularly proud, due to its highly successful results. We also take advantage of all the opportunities offered by our partners. For example, the IC manufacturers train and educate the software staff. Equipment-producing companies offer training for CAD operation. For the management training of qualified engineers, we make use of appropriate international management training organizations.

Aspray: Do you have other comments you care to make?

Bruch: I grew up with this company. It is a fascinating job because our products are designed to entertain people. Consumer electronics is a highly competitive and fast-moving industry. You cannot sit around waiting and planning exactly what is going to happen in the next half year. There are major challenges and there are extremely good opportunities to meet people in different industries. I have contacts with people from unions, universities, and politics. My job requires much travelling, but I enjoy this aspect. I grew up in this area. I love my home and my family and I like my job. It is satisfying that the products we bring to the market are extremely popular in our own country and in Europe. Some products are well known throughout the world. For example, our shortwave world receivers and our dictating machines. You can find them everywhere. It is a pleasure to find a Grundig dealer almost everywhere.

Aspray: Americans are so very concerned about the Japanese these days. What are your experiences of competition with the Japanese?

Bruch: They are our strongest competitors. They are innovative and job oriented. They have a long-term strategy, work rapidly, and think globally. But at the moment the Far East is also confronted with growth-limiting factors. The philosophy of growth has to be adapted to fit in a stable or even a decreasing market. This actual situation requires different methods of management. Perhaps this is an advantage for European companies because they have been struggling in this situation for a long time. But problems did not only come from Asia. There are a lot of homemade problems. Japanese companies recognized very early the importance of quality customer orientation and cost effective production. Overall, I would not say that Japanese companies are more innovative than European ones.

Chapter 2

Ernst Denert

About sd&m

In 1982 Ernst Denert and Ulf Maiborn, formerly chief designer and manager of the START project at Softlab, respectively, founded software design & management (sd&m) in partnership with minicomputer producer Mannesmann Kienzle. sd&m began with DM 600,000 and a software consulting staff of ten employees. It differentiated itself from the German software houses founded in the late 1960s by restricting its functions to consulting activities. A year and a half after its founding Denert and Maiborn purchased Kienzle's shares, thus gaining full ownership in the company. In the early 1990s sd&m expanded its ownership and formed partnerships with the Ernst & Young consulting firm and Bayerische Landesbank.

Since its founding, sd&m has completed software development projects for a diverse array of clients including Lufthansa, AEG, Greenpeace, Siemens, Thyssen, TUI, Deutches Museum, BMW, and Bundesbahn. Its current staff includes 150 employees. The company generated DM 4.2 million in sales in 1992.

Ernst Denert

Place: Munich, Germany

Date: June 29, 1993

Aspray: Why don't we start by having you tell me something about your personal life and career? Maybe you should begin by telling me about the kind of family you grew up in, what your parents did, and what your education was like.

Denert: I came from a lower-middle-class family. My real father died in Russia a few months after my birth. He was an engineer. My stepfather adopted me. He liked me. He was a clerk in several German firms. My mother had no special education, so she was a worker and a housewife. I was born in the Sudetenland, which is in Czechoslovakia. As you know, German people fled from there after the war, after 1945. We came out very late in 1950. We had absolutely nothing when we came out. So my mother had to work. She was a simple worker in a factory.

Aspray: Did your mother and stepfather think that education was important for you?

Denert: Yes. Especially my father. He stressed how important it was for me to make my *Abitur*, which is the degree that we have to have for the university. He stressed that very much and tried to help me, but I soon recognized that he really couldn't help me because he didn't know very much. We lived in Nuremberg, which is 160 kilometers north of Munich. A nice town.

Aspray: Did you have hobbies that were related to science or engineering when you were growing up?

Denert: No, I don't think so. My interests were in playing football. I had an electrical railway but that's not scientific. I also built a radio.

Aspray: Did you show aptitude in math and science at an early age?

Denert: I was in the middle of the school in terms of my grades. I was not very good. But my best results were in physics. I was interested in mathematics, but I had a bad teacher. I was not a young genius. [chuckling]

Aspray: Please tell me about your university.

Denert: I went to the Technical University of Berlin.

Aspray: Why did you choose to go there?

Denert: There were two or three reasons. The first reason was because the Technical University of Berlin was the only technical university having a course of universal studies. We were forced to have not only technical subjects, but also philosophy, linguistics, and history. We had to take four or five subjects from the humanities. Not social science. I wanted to have that course of study so that I would not be so one-dimensional, technically oriented. That was the first reason.

Maybe the real first reason was that in those times the Berlin people did not have to go into the army. [laughter] You could simply avoid going into the army one and a half years. One could win two years because the studies began in the winter semester. Actually, I lost two years. I started my studies two years later because it didn't work. They caught me before I went to Berlin. [laughter] The third reason was that I wanted to go away from home. I have a lot of friends with whom I went to school who stayed in Nuremberg. They are very provincial. I studied electrical engineering.

Aspray: How did you decide to do that?

Denert: I did not feel strongly enough for physics. [chuckling] I had an orientation towards science, especially physics, but I preferred some engineering discipline, electrical engineering.

Aspray: As part of the electrical engineering study, did you study both power and electronics?

Denert: Mainly electronics. Of course a little bit of power is in every electrical engineering curriculum, but it was mainly telecommunications. That brought me to computer science.

Aspray: Were there any faculty members in informatics or computer science at the time you were there?

Denert: No. That was the beginning of my luck in that field. I'm especially lucky that I had to go into the army. Because if I hadn't, I would have finished earlier and then I wouldn't have had the chance to switch to informatics, i.e., computer science. In Germany, they started a massive program in computer science education at the universities in 1970–71. That was exactly the time when I finished my electrical engineering studies. So when I finished they started to build that new department. The first one or two professors came in, of course, in electronics. I had some contact with these things. I started programming and I wrote for a diploma thesis a programming simulation of digital circuits. Discrete simulation, things like that.

Aspray: As part of the regular electronics, had you learned some of the hardware side of computing?

Denert: Some things, but not very much. Switching circuits and these things. We had to work with analog computers, which is a really strange thing. When I received my diploma, I made a very clear decision to go study computer science. I cannot really explain why I studied electrical engineering, but I can explain why I switched over to computer science. I made the choice to pursue a Ph.D. As a "scientific assistant" employed by the university, I had the choice to have a position in the Electrical Engineering Department or the Computer Science Department. I could get the one in Electrical Engineering earlier, but I didn't want it. I waited some months and then it worked to get one in computer science. It was a really big chance because we had nothing. We had no professors in computer science.

Aspray: Were you taking a risk?

Denert: No, it was not a risk. It was an opportunity. We had problems in Berlin getting good professors. They had problems because there was no computer science program before. From where would the professors come? Almost nowhere. In Munich, the computer science education had begun only a few years earlier, perhaps two or three. I think it was in 1967 or 1968. Almost nobody was available. So we had the chance and also, of course, the obligation to study, teach, and do research, because we wanted to earn our Ph.D. degree. Sometimes it was very hard because we had no academic teacher. There were not enough and not good enough academic teachers. So it was a hard time, but a good opportunity.

Aspray: What was the course of study? What were the topics that you defined at the time as appropriate for an informatics program? I know you probably took a strong role in determining what you studied.

Denert: One of my main problems was a lack of mathematics. In engineering you mainly have calculus, not so much algebra. And you have discrete mathematics, which is very important for computer science. I could solve differential equations but did not know algebraic structures. In computer science you have a lot of theory: automata theory, formal languages theory, recursive functions, and things like that. I knew absolutely nothing about these topics, but I had to give some courses together with the professor in these areas that I didn't know. So I had the situation where I was at most one week ahead of the students. That's what I mean when I say it was a hard time.

Aspray: But it is a way to learn a subject very thoroughly.

Denert: Yes, of course, and very quickly. You asked me what is important in my career. I think that the most important event, and the most important period, was 1968. That was the time of the student revolution, which was very strong in Berlin. Berlin was the place where most things were happening in Germany. Frankfurt and Munich were active, but Berlin was the most active. It was the best time of my student life. [chuckling] I was engaged in these things but not in the more political way as people like Rudi Dutschke or others at the Free University. The center of the movement was the Free University.

I was at the Technical University, which was a much calmer place. But, nevertheless, we had an Anti-Vietnam War Congress at the Technical University in 1968. That influenced me very much. A lot of my friends and I were engaged not in generating political activity, but in changing the university policy about the conditions of our studies. For example, there were many regulations regarding the exams. Electrical Engineering was very conservative. One of our main complaints was that it took much too long to finish the course of study—up to fifteen semesters!

Aspray: That's a long time.

Denert: Yes, it was much too long. That had nothing to do with the Vietnam War. [chuckling] But it was important.

Aspray: It was a period of reform.

Denert: Yes, it was a period not of revolution, but of reform. I was very much engaged in that. I think my democratic consciousness stems from that time. Of course we were left wing. There is a basic democratic understanding which I got from this period that influenced me very much. I think that influenced a lot of the corporate culture of sd&m. That is why I am talking about that. I do not think sd&m would be the way it is if I had not been in Berlin in 1968.

Aspray: Elaborate, please.

Denert: Normally big companies are very conservative. I have to do a lot with insurance companies and banks like Thyssen and Deutsche Bundesbahn. These are big companies. They are very, very conservative and they have a culture which does not motivate the kind of people we need to do our projects. In companies such as the ones in Silicon Valley, there is a creative atmosphere with people who are very well educated, who are very engaged, who are very intelligent and who run around in jeans, t-shirts, and jogging boots. They would not feel good wearing a tie. You cannot lead them like a military unit—in an authoritarian way. The 1968 movement was an antiauthoritarian movement. It was a big success, a big event for me when for the first time in addressing a professor in a department session, I didn't say "Herr Professor Cremer." I simply said "Herr Cremer." Do you see the difference?

Aspray: Yes.

Denert: I omitted the word "Professor."

Aspray: The honorific.

Denert: Yes. That was not very easy. It was hard. But it freed us. We had decided to accept an authority only due to its competence, not its position. That was a very, very important experience.

Aspray: Is that something that you practice in sd&m?

Denert: Right.

Aspray: Is it also true at sd&m that the kinds of benefit packages you provide to your employees, or the kind of charities you support, are of a more liberal nature or more left nature?

Denert: I would not describe them as left. But we do have events that go in this direction. For example, in two weeks we will have a summer party at a lake in the west of Munich with all the employees and their families. It will be a very nice event. Not conservative.

Another important example is that we had and continue to have a series of workshops, where we try to bring technical work and social contacts together. Every year we have arranged a workshop where everyone in the company went to a hotel in Austria in the mountains, from Wednesday evening to Sunday. We had two days—Thursday and Friday—with some technical program or lectures on operating systems, or CASE. Sometimes there were lectures on other nontechnical but very important topics, such as how to write good German or on body language. There is a very famous guy, Samy Molcho. He is an actor and does pantomime. He gives seminars in body language. He is a very interesting and amazing guy. So we arranged a workshop for one or two days on body language and on good German because we know how important it is to have a good communication with our customers and within our teams. Our colleagues brought their wives, or husbands, with them and the children. We were together with them in the evening and we climbed a mountain, or did something like that. That is a very typical package of social programs. These things are a very important part of our corporate culture.

Aspray: With regard to this issue you mentioned before of valuing people not for their position, but for their abilities, does this mean that people can advance very rapidly in the company?

Denert: Yes.

Aspray: Does it also mean that people can work more flexible schedules and take care of family concerns and things like that?

Denert: Yes. We don't care when somebody comes or goes. There is almost no regulation. Of course, the project must go on. The work must go on, but we are very liberal in these things. There are a lot of aspects in this regard which I think stem, to a large degree, from my 1968 experience. You must know my partner with whom I founded the company, Ulfried Maiborn. He is one year older than I am. But he does not have that experience. He was an officer in the Army for four years, and then studied electrical engineering at a very small place. We share these values, but the influence of the 1968 culture is more on my side.

Aspray: Let's discuss your career in computer science.

Denert: The problem, as I have already said, was that we had neither enough nor good enough professors. One reason was because they didn't exist at all. But another very important reason was that they didn't want to come to Berlin—to the revolutionary students. They were afraid. It was really a problem. That forced us—a group of relatively young assistants—to take more responsibility than is normally the case when there is an established group of professors, as it is at the Technical University of Munich where I am now. It is unthinkable that a young assistant would take the responsibility we had in those days. I think it is impossible anywhere today. Times have changed. From my point of view, we had the times of the gold diggers in Sacramento. Something of that sort.

We developed a curriculum from scratch for the first four semesters of the computer science program which was very good. I think we were quite ahead of all the other universities in building a very good curriculum. We didn't only have to plan it, we also had to realize it. I had to give one very important lecture. It dealt with data structures, algorithms on complex data structures. Preparation for that led me to write my first book. I had at least a manuscript of the first book before I had my doctoral thesis. Simply because we had to do it. There was only the book of Knuth on the *Art of Computer Programming*, volumes one and three.

Aspray: I know those books.

Denert: Yes. They were the only books we had on these topics. These are very good books from Knuth, but as you know it is not good enough for educational use. So we made our own manuscript for the students. We made it a second time, and a third time. And then it was a book.

Aspray: What was the title?

Denert: Data Structures. Datenstrukturen.

Aspray: Who were the authors?

Denert: My colleague, Reinhold Franck, who is now dead, and I. We used Algol 68 as a programming language to explain the data structures and algorithms. As you know, Algol 68 was unsuccessful in the sense that it was never widely used and is not used today. But it was a very good language for understanding and explaining a lot of concepts.

Aspray: Was it used in some of the universities for a while?

Denert: Yes. We sold about four thousand copies, which is rather good in Germany for such a book. I know some universities used it—not only Berlin. I think the book was a success. It was a pity that, at the same time, Niklaus Wirth wrote a book on data structures. We were unknown, and he was already very well known at this time. His book is a good book too.

I learned, I taught informatics, and we did some sort of research in order to get our Ph.D., which was very hard. But it worked. We were forced to lecture because there were no professors. That gave me the chance to write this book. It brought me forward.

Aspray: Did this book constitute your own research as well?

Denert: It had something to do with my research, but that is not the most important point. The important fact is that I learned to be self-motivated. I simply had to do these things. I had to lecture. I had to develop a manuscript two years after my diploma.

Aspray: What was your research on for your degree?

Denert: I developed it with two other colleagues, Reinhold Franck and Wolfgang Steng. It was a two-dimensional programming language. It is very, very strange, as are most Ph.D. theses.

Aspray: What's the rationale for having one?

Denert: A two-dimensional system means you don't write a program, you draw it. Let's have a look at some pictures here. If you deal with complex data structures, you would like to draw pictures like this. You have data elements which are connected by pointers and you are considering how an algorithm changes that situation. For example, you see the dashed lines here? That means before applying an algorithm to that data structure it looks like this. Afterwards it looks like this. The simple idea was to take this picture as a program. You simply write the data structure in a before-image and an after-image. You see the transformation. That's the effect of the algorithm. We developed a programming language where you can draw programs from manipulating such data structures. We did it in all consequences. We made a formal definition of syntax and semantics and Reinhold Franck expanded the theory of syntax analysis from the one-dimensional string case to the two-dimensional case.

Aspray: Did this catch on at all?

Denert: No. It makes no sense. Quite simply it makes no sense to draw programs. It only made sense because we got our Ph.D. for it. [laughter]

Aspray: Before going on to the next stage of your career, let me ask one last question. In this very special environment in Berlin, were there other people who were affected by it as much as you? Did your cohort group of students have a set of careers so far that were unusual compared to maybe another set of students at some other institution, some other time?

Denert: That's hard to say because normally you only know about the people you knew as a student or as an assistant. But you don't know what happened to the other students. A lot of my friends made good careers in academics. They became professors.

Aspray: But nothing special comes to mind?

Denert: Nothing special, no. In 1976 I left the Technical University, half a year after my doctor's exam, and came to Munich, to Softlab, which is one of the renowned German software houses.

Aspray: Was it renowned by that time?

Denert: It was five years old. They had a good name.

Aspray: What were the options you had when you were looking for a job?

Denert: It was a bad time. Frankly, there was no option. In 1975 we had a recession in Germany. In 1975–76 the economic situation was as we have it now. Maybe not so bad as now, but we had similar problems. It was not so easy for me. I did not have five contracts to choose amongst. I simply had one from Softlab. It was a very good one. Maybe three years later things would have been quite another way.

Aspray: What were you hired to do?

Denert: Softlab made and makes software development projects, and I was hired to work within such projects. That was lucky for me. In these days, Softlab had acquired the project START. I mentioned it in my book, *Software Engineering*. It is a system for the travel market, for travel offices. Maybe you know Amadeus or in the United States you have the Apollo and Sabre systems, though they are more airline reservation systems. START is not a reservation system. It is a big network and a big software application which has to do with the travel business.

Aspray: So it's something that is used by the agents at a travel company?

Denert: Right. On the one side you have the travel agencies with offices in the towns—nowadays there are some ten thousand of them. On the other side you have the providers. They offer travel: Lufthansa, Bundesbahn, Tourist Union International, and all these organizations where you can book a warm-water trip, for example, to Majorca. That system brings those two groups together. It was a big project. It was a big network. It was done by Siemens for START, which was a company founded by

Lufthansa, Deutsche Bundesbahn, and TUI—Tourist Union International. At Softlab we had the application software to develop as a subcontract. It was a project that turned out to require more than 100 man years.

Aspray: It wasn't planned that way?

Denert: No, it wasn't planned that way at the beginning. [chuckling] I think the first estimate was 144 man-months, which comes out to 12 man years. That is typical of software development life. [chuckling]

Aspray: Yes.

Denert: What is important is that I met my later partner there, with whom I founded sd&m. He was my boss at that time. He hired me. He had been at Softlab since 1972, almost since the beginning. He had acquired that project, and we started the project with a small team. At the beginning, I think there were four people. Writing specifications, doing design, all these things.

Aspray: You were in from the beginning of this development?

Denert: Yes, I was in from the beginning. It was a big chance for me. You know, a gold digger? [chuckling]

Aspray: Yes.

Denert: It was another kind of gold-digging that I could do there. Now comes in a very important point—Ulf Maiborn is a very tough, very strong manager. I was always a designer, a software engineer, a technician. There is a famous book by Fred Brooks, *The Mythical Man Month.* Maybe you know it?

Aspray: Yes.

Denert: There is a chapter in it on design and management. Brooks says that it's a good idea to separate these roles of the manager and designer in a big project. He describes the IBM System/360 project, which he managed. We established this separation of roles in that project in the following way: Maiborn was the manager of the project and I was the chief designer. It wasn't defined in that way from the beginning because I was a youngster in the company. But it turned out that after one or two years there I really had that role. It is not by chance that sd&m was called "Software Design and Management." It stems from this experience. We were two strong persons in those two areas. We managed to make that project successful, which at the peak time involved about thirty people.

Aspray: In sd&m, the division of labor remained the same way as it had been on this project?

Denert: Yes, of course. The division of design and management remained. I did not have the chance at sd&m to do only design. We had to manage a company together. But his focus was mainly on the management of the project. My focus was more on the technical side of the projects. It was a big chance. This special sort of gold-digging in the START project was that I had the chance to bring a lot of ideas that I had gained from my university career into that project. I always thought about it and I still do today.

I continue to read articles and think about how I can use this knowledge in the practical life of my project. I sit in the train or in the airplane, read *Communications of the ACM,* or the *IEEE Transactions on Software Engineering,* and I think, "what can I do with that idea? Can I use it?" So I used several interesting concepts for the first time in that project. Then we used them again and again. The most important thing, of which I am very proud, was that we did the START design with the rather new idea of data abstraction. We had to do some sort of design and nobody knew how to do it. At Softlab they had only learned structured programming. Dijkstra wrote his famous paper, "Go to Considered Harmful," and invented the structured programming concept in 1968 and 1972. Then the Nassi/Schneiderman charts came out. Softlab picked them up. It was the big story of Softlab, but it was very simply the use of structured programming. I thought it was not the most important thing. I had the opportunity to meet Dave Parnas. Do you know him?

Aspray: I know the name.

Denert: I had the chance to meet Dave Parnas in 1972 in Berlin. As I told you, we tried to get professors and it was very hard. So we tried to find them in the United States. In 1972 I made a trip to a conference in the States and visited some people, who I tried to acquire for Berlin, [laughter] which was silly. As a young man I tried to pick up a professor in the United States! Dave Parnas was one of them. I didn't visit him in the States, but somehow I came into contact with him and we had him for a lecture in Berlin. It was exactly the time when he wrote his very famous paper on "Criteria to be used in decomposing systems into modules." That paper and some other papers contained the idea of data abstraction as a modularization principle.

At the same time, while at the Technical University, when I was writing the first version of the data structures manuscript for the students, I started programming in Simula 67. Now almost everybody knows that Simula was the first object-oriented programming language. So I am one of the oldest object-oriented programmers because I tried all the algorithms in the book on data structures in 1973. At the same time, Dave Parnas's ideas on data abstraction came together in my head. I tried these approaches for two or three years at the university and then came to that START project. Nobody knew how to do it. So it was a big chance, gold-digging. I think we were almost the first groups to design a real system with commercial conditions using the idea of data abstraction. This was in 1977. Of course there were a lot of papers around the scientific community. They were trying these ideas in university courses, but not in a 100 man-year project under commercial conditions. I subsequently learned from a book by Ivar Jacobson that his group did something similar at about the same time in Sweden. Maybe they were earlier than us. But I am not quite sure. I am very proud about that.

Aspray: Did this work well?

Denert: It worked quite well. It still works today. START is a very successful system and company. They are customers of mine again at sd&m. Six or seven years after the system became operational, they showed me how the system evolved. I was afraid that they had maintained it to death, that they had damaged the structure which we had brought in. But they didn't. There were some young guys at START who were Parnas students.

Aspray: Oh!

Denert: Yes, that is curious. Parnas didn't come to Berlin, but instead went to Darmstadt, which is near Frankfurt. He was a professor there for some years. Of course, he influenced his students and two of the students went to START, where they found a system that was designed according to the principles Parnas had established. [chuckling] And they maintained this structure. They are now in responsible positions within START. They are my friends. They keep that system going, and in a few days I will meet with them to discuss how they can introduce more object orientation in their company because they fear they are a little bit behind the time.

Aspray: Back to Softlab and the START project....

Denert: There were some other ideas and concepts which were first tried out in that project, but the data abstraction modularization is by far the most important. So the lesson from that is that I always tried to bring concepts from the scientific world into *practice. Always in real projects. We have never established a department within Softlab or within sd&m responsible for developing methods we call "green table...."*

Aspray: What does that mean?

Denert: The meaning is that it's developed in theory, not in practice. It is not applied, not within a real project, but aside from that. I had always done this type of development—methodological development, tools development—within a project. We always had the philosophy that it must pay back within the same project.

Aspray: That's a hard demand, in a way.

Denert: But it worked. It always did.

Aspray: What happened to you at Softlab? Where did your career take you?

Denert: I started as a developer and then I became project leader of part of the START program. It was a very big project, and we had subdivided it into two main groups. Maiborn was the manager of the whole project. I had a colleague who had another group, and I was project leader. Then I started some other projects which are not so important. I became a department head for a group of about twenty-five people. That was my last position. Then we decided to start to found sd&m.

Aspray: Why did you do that?

Denert: Because Maiborn asked me! That's the fair answer.

Aspray: What couldn't you achieve staying inside Softlab?

Denert: I could have stayed there. Maybe I would be in a good position there now. Yes, I am sure I would be. When I left the university in 1976 I thought that I would return to the university some years later. For five years after leaving the university, I wanted to become a professor. But it was important for me to have practical experience.

I still have the opinion that it's bad that so many of our professors never left the university and do not have practical experience. That's fine for mathematicians. That's fine for people in computer science in the theoretical field, where there is no practical experience to be had. But it is not OK for practical computer science. It is not OK for professors who teach software engineering. We have too many such professors. So I thought I should gain some years of practical experience and then become a professor. That would place me in a good position. The longer I stayed at Softlab, in the industry, the lower my desire to go that way because I recognized that I was able to do very interesting technical work in practice. I could do things which I could never have done at the university. At the university you can only make projects with a few students, and then not during the semester, but only during the vacations.

Aspray: The vacations? The out-of-term times?

Denert: Right. So they tend to be some sort of toy projects, and I could make very good experience in practice. By 1980–81 I still thought about becoming a professor, but with less desire.

Regarding staying at Softlab, I was not quite sure about that. I had a very interesting project at SEL from this time. We built the *Bildschirmtext*. Do you know what that is?

Aspray: No.

Denert: Videotext? It doesn't matter. They never tried it in the States and that's good. But it was an interesting software system. Software and systems in telecommunications projects. It was a project I made with a Softlab team of about fifteen to eighteen people with SEL in Stuttgart. It was a very interesting, very exciting project. I thought about nothing else. [chuckling] I am a project man. I had some other projects in my department, but they were not so important. I really did not think about changing things. It was an exciting time. I was invited as a speaker to conferences. I wrote papers. I had a very good reputation in the scientific community. I liked it. Of course, sometimes we discussed forming our own company. Everybody at a software house occasionally thinks about being an entrepreneur, doing it for himself. It's quite normal.

Aspray: Is that because there is no capital barrier to entry?

Denert: Yes. It's quite easy. Take a sheet of paper and a pencil and you start a business. When you are an employee of a bank, you don't think about founding a bank. [laughter] But being at a software house is different.

Aspray: Why do you think Maiborn wanted to leave and form a company?

Denert: He is the born entrepreneur. I'm not. I'm the born professor, the design and technical guy. I would like to win the Nobel Prize, which would of course be impossible. I would not be good enough, and it's impossible in computer science. He is a manager and wanted to be an entrepreneur. He saw the success of Klaus Nengebauer who founded Softlab. I think he wanted to copy that. So he was looking for a partner because it's hard to do it alone. It's better to do it with somebody else. We were friends. We climbed the mountains and did a lot of things together. We climbed Matterhorn, for example, which brings you close together. [chuckling] We worked together. We were very successful with the START project. He had sometimes talked about this idea of building his own company. But I always thought that I wouldn't like that. Because if I did that, I would have to do much administrative work and acquisition, all the things I didn't like. It would hinder me to do my technical work. So I didn't like that idea. I wanted to get a professorship. I couldn't imagine building a company. We started sd&m in 1982. The decision was made at the end of 1981. In 1980 I couldn't imagine building a company. I thought it would be a big nonsense for me. But things changed. We discussed it. The decision was made in one evening. We were in Stuttgart due to the SEL project. We had a dinner together and some glasses of wine and he said, "Don't you want to? Shouldn't we make that? What do you think about it?" I said, "OK, let's do it."

Aspray: Goodness! Did you have any sense for the kind of company, the kinds of clients, the kinds of work you wanted to do?

Denert: Yes, of course. It was quite clear that we would do a similar type of business to that which we had experienced at Softlab. The similarity is software development projects for clients. No software products. So we got started. We did not need very much capital. Our experience was that software design and management is important for making good projects, and this differentiated us somewhat from the early German software houses, which were founded in the second half of the 1960s.

Aspray: What companies do you have in mind?

Denert: We were thinking of companies like PSI. They are more technically oriented and are located in Berlin. Or SCS, which are now with cap debis. GEI, which was a subsidiary of AEG and is with cap debis now. Or ADV/Orga which is now within Sema Group. Or Ploenzke. They are all well known or got big. Most of these were founded in the 1960s. Softlab was founded in 1971. These software houses earned their money in the 1970s mainly by body selling. They sold programmers to customers. They also tried to make projects because some of them were connected to big companies. mbp, for example—to Hoesch, which is a steel company.

Aspray: GEI was connected with AEG?

Denert: Yes, so they made systems solutions for them. In its earlier years Softlab earned its money through body selling.

Aspray: But you wanted a value added to your project?

Denert: Yes. We felt that it would be not enough in the 1980s to be able to program—only to have good programmers and sell them to some customers. We had experienced, especially from that START project, that software design and management is important for making projects. We didn't want to only do body selling. We wanted to do projects. So from the beginning, sd&m made projects. It took the responsibility for making projects for customers. We prided ourselves on good design and management, as well as on good programming.

Aspray: What did Softlab's management think of the two of you leaving? You were, after all, a competitor to Softlab. Right?

Denert: Yes, of course. We were and we still are competitors. Of course they didn't like it. But they had to accept it, there was no choice. In the first years I think there was not very much friendship. Especially between Maiborn and Nengebauer. We had no contacts. We met our former colleagues, old friends, of course. But we didn't steal people. We were quite fair. They tended to accept us as competitors.

Aspray: I was just about to ask a slightly more general question as to what you felt was the ethical way to behave towards them? Obviously you believe not taking people from their operation was important. What other sorts of things? How did you deal with customers, for example?

Denert: We tried to get a project that we started at Softlab. It had to do with Deutsche Bundesbank. The early phase of the project had been finished, and when they asked Softlab for an offer for the next phase they also asked us. We made them believe that they should ask us too. But we lost it. So we took no clients away. There is another important point I shouldn't forget. We started sd&m not only with two private owners but also with an industrial partner. Do you know Kienzle?

Aspray: No, I don't think so.

Denert: They make minicomputers. They are smaller, but can be compared with Nixdorf. They are like Nixdorf, but not as good. They were bought by Mannesmann. Maybe you know Mannesmann?

Aspray: Yes, that name is familiar.

Denert: Mannesmann Kienzle is the company with whom we started sd&m. They had 50 percent at the beginning and we each had 25 percent. They thought they would go into bigger software projects, so they needed somebody who could do these projects, and they thought we could. But it turned out that they went on to do their minicomputer business—selling minicomputers to middle-sized companies, just as Nixdorf did. They did not take on big software projects. It was not their business. They realized it after one or one and a half years. We also saw it. Management changed there. After one and a half years we bought their shares, and they were out. But it was very helpful to start with them. Because for the first year, sd&m's work was done for Kienzle.

Aspray: It gave you projects to have to work on.

Denert: They gave us projects that were not very good ones, but it was a good start. It gave us the possibility to start not only with two employees, but with a team of ten people. The two of us and eight other people. A secretary and developers. That would not have been possible without them. After one year had passed, we had almost no more projects from them, but instead we had customers such as Lufthansa, TUI—which is Europe's biggest broker of packaged tours, and Bertelsmann. These were very renowned customers.

Aspray: Had you known these companies from Softlab?

Denert: Yes. Lufthansa and TUI were involved in START, of course.

Aspray: Had you also had some other expectations from Kienzle? For example, did you expect them to provide certain kinds of administrative functions for you?

Denert: No.

Aspray: Or provide introductions to banks? Or any of those things that are done in running a business?

Denert: No. The main expectation was that they would develop their business with more software orientation—that they would provide systems solutions with packages of software and hardware. At the beginning our work had to do with that project which we tried to take over from Softlab with Deutsche Bundesbank. Deutsche Bundesbank had a project with Kienzle where we supported Kienzle from Softlab. This was part of the connection to Kienzle. Kienzle believed that they would have other projects like this one for Deutsche Bundesbank, with other customers too. But that didn't work. They lost the

Bundesbank project, and so they went back to their original business, selling boxes and providing solutions for small and middle-sized companies. That was good for them. It was a dream that we dreamed together with them, with the manager of Kienzle. But he was thrown out and another manager came in.

Aspray: Did you have difficulty convincing Kienzle? Did you consider other companies when you were looking for your industrial partner?

Denert: We didn't look for other partners. We were not in the position of going around in the industry and saying "Hey, here we are, Maiborn and Denert, we are looking for partners." It was a personal connection mainly between Maiborn and that manager from Kienzle, Dr. Bindels, which stemmed from that joint Bundesbank project. It is always a matter of personal relationships, whether such things work or not.

Aspray: But you didn't have serious problems convincing Kienzle to enter into this arrangement? It sounds like they went quite willingly into this.

Denert: Yes, they did because that one man wanted it. He had the vision that Kienzle would have more of these kinds of projects.

Aspray: Was Kienzle a publicly held company? Or was this private?

Denert: No, it was held by two brothers named Kienzle. It was a family company up to 1981. Then the first 50 percent was bought by Mannesmann, which is publicly held. Then 100 percent.

Aspray: So the decision didn't have to be made by an overseer board. It was a small company that worked in-house.

Denert: Yes.

Aspray: Now we come to the breakup of the two.

Denert: Yes. They saw that the connection got looser, that the two companies didn't work very much together, especially in this project business. It was Maiborn's part. I couldn't believe that they would like to get out because we were profitable in our first whole year of business. Already we made good profits.

Aspray: That's really very surprising.

Denert: Yes. We started in October 1982. Of course, that first partial year we had a starting loss. In 1983 we had profit. At this point in time Kienzle went out. But they could see that we made a good development. I couldn't believe that they would like to get out, but it wasn't their business. They didn't want to be the owners of a small company making a profit. The management people were afraid that we could make some nonsense. They did not have the majority. They had 50 percent ownership. They knew that they couldn't really control us. They were afraid that we could make some nonsense. Here was a company with a big name—Mannesmann—who would be made responsible for the nonsense we might make.

Aspray: So there was more risk than profit for Mannesmann.

Denert: Yes. The profit they could expect from us was not worth the risk they took. They were willing to get out. It was a very fair agreement. We bought their shares, and then we were free.

Aspray: Was it difficult to be able to buy those shares though?

Denert: No. It was a very fair price. We had already made some profit here, and so it could be done. We could finance it. It was absolutely no problem.

Aspray: What other early challenges did you have in that first year and a half that you were together?

Denert: I think the main challenge was working with Kienzle. Of course in the first year 100 percent of our contracts came from Kienzle. Although the management there liked us (otherwise they wouldn't have made the deal with us) people with whom we had to work were afraid that we couldn't take over their work. You must remember a very important point, that in 1981–82 we again had a recession in Germany. The first recession I experienced was in 1975, when I tried to get a job. The second was when we started this company. [chuckling] But it was a good time for us because we were in the economic valley, and after that it went up. It was a really good starting point in this respect. But of course, it was a hard time and people were afraid of losing their jobs. There we came in at Kienzle, doing work for them. So their employees didn't like us. A lot of people with whom we had to work made things difficult for us.

Aspray: Were any of the employees you took on as developers from the Kienzle staff?

Denert: No. We took our employees, our developers from the market. We hired them, which was not very hard because it was a recession time.

Aspray: It makes it easy. It was a buyer's market.

Denert: Yes, of course. It was a strange situation for those people whom we hired, because we conducted the interviews in our home. We had no office. [chuckling]

Aspray: Were you working out of your home entirely?

Denert: No. When we started work on the first of October 1982, we had an office, but we had to hire the people some months earlier.

Aspray: I see.

Denert: It was a time when we still were working for Softlab.

Aspray: So you are set off on your own without having the safety net of Kienzle there to provide you with work. How optimistic were you? What was your plan for developing your own client list?

Denert: We were optimistic about getting our own customers because we had contacts with Lufthansa, TUI, Bertelsmann. We approached the Deutsche Bundesbahn, which is our most important customer when I look over all the years. We were very optimistic that we could get some customers. I know it sounds strange, but I never had the feeling I was taking a very high risk. Normally people believe one takes a high risk in founding a company. But I always thought, if it doesn't work you can be an employee at a good company, at a software house as a software project man. We started the company when I was forty years old. I was not afraid about my situation.

Aspray: Had you invested a lot of personal savings, though, into the company that would have been lost?

Denert: Yes, but it would have been only the past which I could lose. It is quite simple. We started with 200,000 Deutsche Marks. But what we really needed was to have that upgraded to 600,000 Deutsche Marks. Where Kienzle had to pay 50 percent, 300,000, and that meant for me, 150,000. I had 50,000 in my savings. And I got the difference from the bank. I got 100,000 marks from the bank. That was my risk.

Aspray: It was significant, but not overwhelming.

Denert: No, it was not overwhelming. We never accepted a debt which could not be paid back. You can earn enough money as an employee so that you can pay back 100,000 Deutsche Marks in ten years. It is not an overwhelming problem. We got to know a colleague of Maiborn who had founded a similar company on robotics. He had taken a debt of two million marks in order to make an investment in a robot or something like that. He had two million marks and the company went down. The company was liquidated and he ended up with a debt of about two million marks. You are never able to pay back this amount as an employee. That means the bank will take all the money you earn up to the minimum you need for living. So it makes no sense and it is no fun to work. We never took such a risk. I always said, the worst that can happen is I lose my past—the savings. But I never lose the future. This other man lost his future.

Aspray: The trade-off by taking that approach is that you cannot grow very fast. You do not have the capital to grow fast.

Denert: We did not want to grow very fast. For some time we had the philosophy that we did not want to grow beyond fifty people. Nowadays we are not very big. One hundred fifty people is not very much. I would say these companies like SCS, Ploenzke have or had five hundred to one thousand people. So we are medium size. We did not want to get very big because we knew and know that we would lose our competence. The level of competence would sink if we were very big. So we always tried to avoid that. So we did not need much capital. If you are in a consulting business, you don't need very much capital.

When you are developing products, you have to make investments in order to develop a product which you then hope to sell. That is quite another story. We never did that because it was not our business. I don't want to say that we did not want to take the risk for that, but it was not our job. We were people making projects. We were staying small or staying the size which allows us to maintain quality. That is very, very important.

Aspray: Do you think that differentiates you from your competition?

Denert: I think so. There are others who may have a similar philosophy. But look at the big players. Let's take Daimler-Benz. They can only think big. They think they can only be successful by buying companies and getting bigger and bigger. Later on we will come to the topic of our partners which we acquired, Ernst & Young. About three years ago, we had contact with a lot of consulting companies in order to check whether they could be our partners. We got a report from cap gemini. Serge Kampff, the big boss of cap gemini, thinks big. I read their figures and they were already bad three years ago. Now they have red figures because they are so big. cap debis is.

Aspray: Can you give me a sense for the size of the market? What is the total dollar value for the kind of work you do for your whole market niche?

Denert: Do you mean how big is the market niche expressed in dollars?

Aspray: Or Deutsche Marks.

Denert: I don't know that. Sorry.

Aspray: How many competitors do you have?

Denert: Well, there are a lot of companies who are sometimes, somewhere, somehow competitors. But there are few which are important. Do you know Software AG? They not only have their database system, Adabas, they also established an application development department. They are competitors. But they are not very good. Nevertheless, they are acceptable from the customers' view. The customers believe that they know the products of Software AG, Adabas, Natural. Maybe so or maybe not. So Ploenzke and Softlab are competitors. cap debis, not so much. There are several others.

Aspray: It's just a handful.

Denert: Yes. That's an interesting point. We often get our contracts without competition. Of course companies must allow others to submit a bid. Especially if they represent some governmental organization. They have to make a public request for bids.

Aspray: They call for offers.

Denert: Yes. They have to have several offers, several competitors, and that's always a bad thing. That's always a prize fight. You don't have a good chance, if you are not already in, as we are at the Deutsche Bundesbahn. Often, we are not allowed to present our offer. You can only write it out. That's terrible. I don't like that. I think we are not very successful in those cases because we are not willing to make low prices. We often get our contracts without competition.

Aspray: Do you often shy away from government contracts?

Denert: Yes. We make these offers, but it's not our main business. We have a lot of contracts, a lot of customers because we have established a network. We have spent a long time in this business world. If you are good, you become well known and you have a lot of personal relationships, and that brings customers to us. That's much more important than a public call for offers.

Aspray: Are decisions about awarding a contract made by these clients mainly on a basis of competence or price sensitivity?

Denert: They are not based upon price. Contracts where we don't have competitors are not price sensitive. Where there is a call for offers, it's very price sensitive.

Aspray: Can you give me a profile of the kinds of companies that are your clients? The ones you mentioned earlier are all very large and quite distinguished companies. Is that the character of most of your clients?

Denert: Yes. I think I gave you a brochure. The brochure lists many of our clients like TUI, Lufthansa, Deutsch Bundesbahn, Thyssen Steel. AEG. IBM is not a very important customer of ours, but it's a good name. The Automobile Club, ADAC, and BMW. We did a small consulting project for them, but it's a good name. That is only a selection. But we have mainly well-known names in there. That's typical.

Aspray: What about the size of the projects that you undertake for them? What's typical?

Denert: That information is in the brochure too. The typical size of a development project—not a consulting project—would be ten man-years. You see the figures here.

Aspray: They range from 10 to 130 there.

Denert: That is not one project. That is a bunch of projects. That represents several projects that sum up to 130 man-years. The asterisk means that this effort is done together with the customer. But we are responsible for the whole project. So we are managing them too.

Aspray: I see. They have some technical staff, some software engineers that work with you.

Denert: Right. They also provide project leaders for parts. But we are responsible for the whole effort. So we think we can count it in this way and say these projects were done by sd&m with their staff.

Aspray: Do you ever not bid on development projects because they are too big or too small?

Denert: No. Nothing is too small. Most things start small and get bigger. [chuckling] And maybe you don't know at the beginning how big it will grow.

Aspray: Are there some projects that are going to be enormous and cause you to hire a lot of extra staff?

Denert: Yes. These projects exist. At the moment we are not asked to make an offer for them. I know two of them. I'm not quite sure whether these projects will be successful. For example, Andersen Consulting makes one of them with the banks. Maybe the companies who make these projects will have to learn that it's not a good idea to give the whole responsibility of such a project to one firm. They started with a contract of about 100 million marks. And now you can read in the newspaper that they are at 300 million marks. Don't quote me on that, but these figures are in the air.

At the moment, we are not in the position where we would be asked to make such a project. People think that we are not strong enough to do it. I think we could do these projects, but we never would take the responsibility for a fixed price, such as 100 million marks. We could not afford that. It would be too much. I also think it's a mistake to do it that way, to give such a large development away as a whole project to one competitor. I think it will not work. They have to learn that.

Aspray: Do you do many projects with other outside contractors?

Denert: No. Typically not.

Aspray: Is your work done on fixed price contracting?

Denert: More and more, yes. When we started ten years ago, we worked on a time and material basis. Now we have fixed prices. We call it *"Werk."* We have a fixed price for a fixed system.

Aspray: Is that a result of maturing and knowing your prices better? Or is that the direction that the marketplace has gone?

Denert: Both things are true. I think the market is changing. And we have learned to live with that. To calculate it. To manage it. If you manage to calculate correctly, it's a good thing because you have much more freedom in staffing and in managing the project. When you work on a time and material basis you always have to ask the customer whether he agrees that you take out this person and bring in another person. That's ugly. I don't have figures at the moment, but I think that more than half of our projects are fixed price projects.

Aspray: Software projects are notorious for growing to be much larger and taking much longer than one expects. How did you learn to manage them? What kinds of lessons did you learn? What were your experiences along those lines?

Denert: We only had one big flop where we made red figures in a project.

Aspray: That's remarkable that you can single out just one project like that. Do you have a philosophy of management for controlling the costs and the timetables for such projects?

Denert: Yes. This afternoon I had a discussion with three of my project leaders regarding how we could make it better. I am sure that sounds strange, but we don't have a very strong system in handling that. We do not have very much bureaucracy and tools and administration and things like that. I would like to have a little bit more of it. [chuckling]

Aspray: Is the nature of your business such that the kinds of development projects you're working on are very much like the ones you've done in the past, making it possible to learn from those previous ones? Or are you always entering new application areas?

Denert: In principle we could learn from past projects. We could gather information by doing a calculation afterwards and compiling numbers from the previous projects, but we don't do it. It's terrible, but we don't do it. Another problem is that we have a lot of young people. We have a very young team. So we have a lot of inexperienced people. They don't bring the experience from a former project. If you estimate the effort it can be terribly false. But as important as making a correct calculation is, if you have one, try to stay within that budget. You can gain a lot of money or save a lot of money if you decide to stay within that budget. Every project consumes its budget. If you double it in the beginning, you will need it anyway. So the art of managing is to stay within that budget. Which, of course, also means to control change requirements. You will always have changes. The customer will want changes. It can be a very dangerous thing when our people, our developers, and the customers' people on the working level agree upon a better way to do it. They decide, "Oh, that would do better, let's do it." So we have effort that was not calculated. It's problems such as these we have to be aware of.

Aspray: Tell me about growth of the company over time. I know you have taken on some partners and you have more clients now. You have grown to 150 people. How did this take place and why the extra partners?

Denert: As I told you, for several years we had the philosophy not to grow beyond the 50-person limit. Now we are 150. We didn't want to grow very much in order to maintain our quality. On the other side, we must have a certain amount of people in order to be able to do big projects. In order to reallocate people and to react to requests from the market, so we need a certain size.

Aspray: Is it also necessary to have people with different specialty skills? Or is one good programmer equivalent to the next good programmer?

Denert: I believe that it's not very important that somebody knows a special programming language or a special system. Customers often believe that. It's sometimes a problem to explain to them that is not the case.

Aspray: Are people sometimes better at one computer science skill than another?

Denert: Yes, of course.

Aspray: So you need some skills of different types.

Denert: Yes. And if you have more people, then it's easier to have a good mix of people. So we didn't really decide to grow further, but it simply happened because we were good. We had contracts and projects and things went on. For the first seven years, up to 1989, sd&m was a two-man show. Maiborn and me. And then project leaders. Nothing else. We founded our first office outside Munich in Hamburg in 1988. Half a year later we opened another in Duisburg. It had to do with a project with Thyssen, because they were located there. We thought that it would be worthwhile to come closer to customers, of course. It was one reason for growth. Again that was mainly Maiborn's idea.

Aspray: Was it his plan?

Denert: Yes. We always discussed it. How could we broaden our management team, our top team? We had the problem, and we still have it, that we are very technologically oriented. We must be that way. That means that we are attractive to technically oriented people, but not so much to managerial people. That made it hard to find the people within our team who could take more responsibility and assume management positions. But we saw that it had to happen. Then we started to take other people in. We also needed that because we had some conflicts. I think it's unavoidable. I already mentioned something to you. It's unavoidable that two people who work closely together for a long period, ever since 1976, will run into some problems. Then it helps when other people on a similar level come in. Then you have someone in some sense between us, which helped.

Aspray: An intermediary?

Denert: Yes. We recognized that we needed to broaden our top team and we did it. When it was only Maiborn and me, it was not always quite clear who was responsible for what. That was one of our problems. We changed it by taking another guy in on this level and establishing other divisions. We now have a structure which enables us to manage this team of 150 people, which was not possible in the previous structure.

Aspray: So in the second version, do some of the divisions report to one person and others report to another person?

Denert: Yes.

Aspray: So essentially what you're doing is creating three, small, fifty people companies again.

Denert: That's the idea. That was necessary in order to manage that growth. The second thing that happened during the last two or three years was a search for partners. Without partners we relied totally on the network comprised only of the two of us. We each had our own network. And that is potentially dangerous.

Aspray: So any projects that were brought in were because you or your partner knew the people that were involved.

Denert: Right.

Aspray: So it's difficult if you want to retire or if you are going to grow and want more customers.

Denert: Right. And at that time my coauthor, with whom I did my Ph.D. thesis, Reinhold Franck, had an accident in the Alps. We both knew him very well. He suddenly fell into a gap of a glacier some thirty meters deep and was dead.

Aspray: It probably made you think about what would happen.

Denert: Yes. So the idea was to look for partners. It only works when they are shareholders. It makes no sense to have some sort of cooperation contract. It doesn't work in the long run.

The basic idea was to get partners in who can enlarge our network and make our position in the market more secure. Then the company would be standing on more columns. One idea was to have a company as partner that was already in the consulting business. But it would make no sense to have one of these companies, such as cap debis, as a partner because they are doing the same business. We had a lot of requests from companies from the United Kingdom, from the Netherlands, from France, who asked whether they could buy shares because they all wanted to move into the German market. The Common Market beginning in 1993 was the motivation for them to look for companies where they could broaden their market share. But it made no sense for us.

Aspray: It was just increasing your competition in a way. So you were looking for a company that does management consulting or accounting consulting?

Denert: Right. A company of that type adds value to our marketing without adding competition. We looked at a lot of consulting companies like McKinsey and Diebold. We ended up with Ernst & Young. I tell this story to many people and in Germany there are a lot of people who don't know what Ernst & Young is.

Aspray: But in America they are very well known.

Denert: Our second partner is a bank, Bayerische Landesbank. It has to do with the savings and loans banks. They are very widely distributed. There's an old tradition of poor people saving their money there. So they are everywhere. There is one in every village. You may not have a Deutsche Bank there, but you have a savings and loan bank there. And Bayerische Landesbank is the head organization in Bavaria. Bayerische Landesbank is the second largest one in Germany. This whole organization of German banks is not one company but it has some loose unifying connection. When considered as a whole, they make up the biggest bank in the world.

Aspray: I didn't know that.

Denert: And then there comes some Japanese banks. We never had to do business with banks and with insurance companies previously. Rather, I should say we hadn't had them as clients. We only had some small projects, but nothing worthwhile. That's not good because nowadays banks and insurance companies are very important data processing organizations. So the advantage of working with the banks and insurance companies is the opportunity to go into that field. We make projects for Bayerische Landesbank and they work much better than they did for Kienzle ten years ago. We learn a lot about banking and we also try to enter the insurance field, which is quite another business.

Aspray: In a sense, banks don't compete with one another the way that other kinds of companies in the same industry compete.

Denert: That's right.

Aspray: So when you learn about dealing with the Bayerische Landesbank—the business—does that mean you can go to the other states and go to their landesbanks?

Denert: Yes, there are some who are friends. Others are more opposed. But of course I get recommendations to visit the Landesbank in Frankfurt or in Hamburg or elsewhere.

Aspray: But if you had been doing work for Kienzle, they wouldn't want you to turn around and do work for Nixdorf.

Denert: They wouldn't have hindered it. But it wouldn't have worked quite well, that's right.

Aspray: But there doesn't seem to be the same kind of problem in the banking industry.

Denert: No. I don't think that is so. It depends on the individual. I met a man at the Deutsche Bank who was angry about the fact that I hadn't told him at the beginning of the talk about our connection with Bayerische Landesbank. I couldn't say a word because he talked so much. I think that is rather seldom. Normally that's not a problem that we have that connection.

Aspray: So you believe that this will be a growth industry for you?

Denert: We are building one of the business areas which I am leading at the moment. We call it financial information systems. Currently it's a rather small team of about twenty-five people. We think that it's a growing field so I'm especially taking care of this one. The insurance companies have to do a lot of projects in the near future. I see that because of the deregulation of the European market. They all have to change their business. They call it "re-engineering." What a word!

Aspray: Did you go to the Landesbank for other reasons? For example, did you need capital, or did they have financial management experience that you didn't have?

Denert: No. We went mainly for this market opportunity. Nothing else. You see, we don't need much capital. We have a capital of about 4 million marks, which is necessary to keep the business going, simply stated. We do our work, pay our people, and get our bills paid three months later. There's a gap, so we have to have the capital to bridge it. But that's all.

Aspray: You still haven't gotten into a position where you have really significant capital expenses. You are not having to buy powerful computing hardware and you don't have investment in plants of other sorts.

Denert: Of course we have a lot of small computers. They get cheaper and cheaper, as you know. PCs and Unix systems are small systems which we use as development work benches. We don't need to own the type of hardware our customers own. We use our customers' systems of course. We are connected to their computers. We have our PCs and Unix-based development environment and a connection to the mainframe. We are transferring the programs we are developing to their computer and translating and testing them there. So we don't need to buy the type of computer they have. That would be impossible. We cannot have an IBM mainframe and be as based with all the system software having IMS and DB2, et cetera. It would be impossible and makes no sense. We do not need very much capital for that reason.

Aspray: This third person you brought in at the top, was this somebody from within the company or from outside?

Denert: From outside. It's a former colleague from the START project. We knew him from Softlab. Meanwhile, there are other people too. So we broadened it on two levels.

Aspray: Have you found that as you have grown, your company has become more complex? Have you had to bring in professional business skills and external people—a very good marketing person, a very good financial person, et cetera?

Denert: The best marketing and financial person is Maiborn. [chuckling]

Aspray: He has the skills already and you didn't have to look for them elsewhere.

Denert: Yes. We are looking for those skills because we need more of them. The third person we brought in didn't bring this special skill, but he brought the ability to manage projects, to manage these division heads, and to manage the projects beneath them. It is his special skill to work with all these people. He made a very important development. We tried to get people on this level from outside, but we failed.

Aspray: I see. The division heads.

Denert: We had some bad experiences. We hired some people who are no longer there.

Aspray: Why do you think it failed?

Denert: They were the wrong people. It is very hard to get good people on this level who fulfill our requirements and who fit into our corporate culture. I hired one in the area of financial information systems and I am very satisfied. He has been with us two or three months now.

Aspray: Isn't it a little early to tell?

Denert: It's not definite. But I think it will be a success. But this task is very hard. The people who are good for doing our type of project work are very technically oriented and have a lack of managerial skill.

Aspray: And interest?

Denert: And interest, right.

Aspray: What about the project staff? What do you look for when you are hiring them? Is it hard to get good people?

Denert: Absolutely not. It is easy. It is especially easy for me to get young people. We take mostly young people. They come from the university—computer scientists, mathematicians. About 50 percent of our people are from informatics, computer science. Between 20 and 25 percent are mathematicians.

Aspray: What kind of mathematical training?

Denert: All sorts. They have some sort of programming experience.

Aspray: Was there something in particular you were looking for there?

Denert: No. They must be good, that's all. There are physicists, engineers. Almost no people from economics, which is astonishing because....

Aspray: So much of your business is application in these areas.

Denert: Yes. But economics students are not very well trained in logical, abstract, systematic thinking. [chuckling] It is a problem. It is easy to understand economics, but it is hard to train students from economics in logical thinking. You can't do that. It is easy for us to get good, young people because we have a good reputation. Where we are known, we have a good reputation. I'm very active in the university. I have a lot of students. My book on *Software Engineering* helps us to get well known. People read that book or hear my lectures and think "I would like to do project work in that way." Therefore they want to come to us. That's the main reason.

Aspray: Do you hire experienced programmers very often?

Denert: No. Not very often. We would like to do it. We have tried it, but we are not very often successful.

Aspray: Because other places keep the good people?

Denert: Yes, and because sometimes the experience doesn't fit our requirements. It's hard to get good, experienced people who fit our requirements.

Aspray: They have learned a different way of doing things at another company, perhaps?

Denert: Maybe. They are rare anyway. Maybe they don't want to move. If they are really good, they may be in management positions and are no longer programmers. So it is not very often that we get experienced developers.

Aspray: Do most of your young people come from Munich? Or all over Germany?

Denert: A lot of them come from Munich, of course. But also from other places, Karlsruhe or Erlangen. From many different places.

Aspray: Are there other things you might want to say about your relations with customers? How you continue them and how you build them up?

Denert: A lot of customer relationship stems from our technical reputation in software engineering. This is not always good. I don't like that. I would prefer to be hired because people need some project done, not because we have a special knowledge, a special know-how in software engineering and in being a consultant in software engineering.

Aspray: I see. So they want to appropriate that special way of doing things and then say goodbye to you?

Denert: Yes. That's one problem. But on the other side it helps coming in. Today I had such a talk with a customer. It was interesting and typical. There is a large insurance company, Aachener and Munchener insurance, the second biggest German insurance company. It's not one company, it's a conglomerate. We call it a "concern." Connected companies. There is a holding organization and then there are different individual companies.

One of these companies made a project and they developed a software architecture according to my book. They had a consultant from another company who had read my book who had told them make it like Denert has written it. Somehow I was asked to make an evaluation of what they had done. They had really done a good job and that gave me the opportunity to present the result of that evaluation, not only to that bond company, but to half a dozen of the others in their concern. They brought together all their data processing leaders and the level of management above them also. One of them asked me to visit him and make a presentation. He has an office nearby our office in Munich. I was there today and we discussed how they could take over this architecture, what they could do with that and how we could help them. That is a typical way of establishing a new relationship to a customer.

Aspray: You are essentially in two businesses, as I can see it. You are in the software development business, but you are also in a consulting business.

Denert: Yes, you cannot differentiate them.

Aspray: Does the consulting business have a strategic value to your company? Is that one of those ways of bringing in new customers? You start as a consultant and then you become the developer?

Denert: That's what I want. It works, but not every time.

Aspray: Is it also the way that you can continue relations with companies after a project is completed? Who does maintenance, for example, on projects that you've developed?

Denert: There are two possibilities. They do it or we do it. We like both cases. In the first case, when they do it, it means that we developed the system that they are able to maintain. We did a good job. We could pass it over. In the other case, when we do the maintenance, it is also a good thing because we have an ongoing contract.

Aspray: Is that a significant part of your revenue?

Denert: No. There are two or three cases of this. Mainly we pass the system over to the customer.

Aspray: What about training and continuing education of your employees? What role does that play in your business?

Denert: An important role. I tell applicants for jobs that the best education is on the job. Ninety percent of what they will have learned when they leave five years later, they will have learned in the projects they have done. I think we have a really good team. Everybody can learn a lot by working together with people in this team. That's the main thing. We provide training within the project if there is a new programming language or system to be used. I think we have a good library. We hold these workshops I already mentioned, on technical and nontechnical topics—interpersonal communication, for example. Furthermore, we hold meetings nine to ten times a year, on Friday afternoons. People can attend external seminars. But the main thing is the daily work.

Aspray: Is there any value in formal course work at the universities? Do your employees find anything that they can't learn better on the job?

Denert: Well, they have done it.

Aspray: Is there a pattern of people going back and enrolling for advanced degrees or additional course work?

Denert: That is not usual in Germany. I'm not sure whether it would be worthwhile. I don't believe so, but I would have to think about it more seriously. But it isn't established, it isn't usual.

Aspray: What about people going back for business training, business courses?

Denert: Business schools?

Aspray: Yes.

Denert: It is also unusual.

Aspray: Well, none of your people want to be managers.

Denert: Yes. The business schools don't play a role with us. Nobody from such a school works for us. This is not to say that they are bad. But it simply doesn't happen.

Aspray: Can you please talk about your views of what happened in software methodology in the 1970s and the 1980s, the so-called software crisis and all of the formal approaches that occurred in the 1980s?

Denert: The software crisis.

Aspray: Do you think there was such a thing? Some people doubt that there was.

Denert: Yes. At least everybody talked about it. In some sense we still have it. We really haven't learned how to build software in an engineered fashion. At least we haven't learned it on a broad basis. I believe I know it and some other people also know it. It is quite another question whether all our people at sd&m really know it. I hesitate to say that they all know it. It is not a very broad skill. Not all the organizations have the skill to really make good software. They have some good people and some companies are better than others of course. But most of them are not really good. They didn't really learn it. We had a software crisis and still have it because people really don't understand how to make good software—or they don't do it if they know how. This leads to systems that are not good. They work somehow, but they do not work really well.

The second problem is that the people who are responsible for selecting methods and tools and managing development departments cannot really judge what is a good method, what is a good process model, what is a good way of processing, what are good tools. The CASE market makes people believe that using their methods will help to overcome the crisis. But that doesn't really help. That's my opinion.

Aspray: What does it do? What would help? Why don't methods such as CASE help?

Denert: It is a strange thing for me to see that some methods—especially structured analysis and structured design and the CASE tools—are not really a good methodology. I question whether it's a methodology at all. I told you that I came from that thinking school of Parnas and data abstraction, Simula, and things like that. There was a totally other school established by people such as Ed Yourdon, Larry Constantine, Tom De Marco, and James Martin especially. I think it was totally misleading. It had nothing to do with the school that I liked—with which I was successful in doing my design. I don't understand why that happened. But it is a fact that it happened. It was supported by glittering CASE tools, very nice looking CASE tools, with a tremendous marketing success. Many people believed that they were on the right track if they bought a tool with such marketing success, even if they didn't understand what was going on. It is strange. That holds at least for the area that I'm in, namely business information systems. It has not held so much for technical software. But this whole bunch of commercial software is done in a very bad manner and there is no understanding of good software here.

Aspray: Taking you back a step or two, why do you think the timing was such as it was for this so-called software crisis? Presumably there were software problems all along. All of a sudden you heard about it everywhere you turned.

Denert: Yes, but that was in the 1970s. I think it is quite simple. The reason was that we had learned to write small programs. People like Dijkstra taught us programming in the small. Nobody told us about programming in the large. Only a few people, notably Parnas, went in this direction. But it's very hard to teach programming in the large. I tried it at the university, but maybe it is not teachable.

Aspray: The whole structure of the university doesn't lend itself to that.

Denert: Maybe other engineering subject areas have a similar problem. I think universities don't really teach how to build a large building, or how to bring a rocket to the moon. It's an experience that has to develop in practice. Maybe we cannot demand it from the university. So having learned only programming in the small we had to do programming in the large. We had and have to do it. That's the reason for the crisis.

Aspray: I wondered if the crisis was also a byproduct of the professionalization of software—and of computer science more generally. All of a sudden, at this time, you have all these universities that are offering degrees and they have students to place and such. And they want to show that there is a need for the product you're putting out.

Denert: So you think they invented the software crisis in order to establish their market? [chuckling]

Aspray: Exactly. It's a conjecture, I don't hold to it necessarily.

Denert: Maybe. People no longer speak about a software crisis. That's a term which is out.

Aspray: The 1980s seem to be rife with all these formal approaches to resolve problems of software productivity. You suggest in the introduction to your book that the 1990s will be a reconciling of the 1970s and 1980s in a way. But balancing approaches and finding a new place for formal methods is not the solution to things. Do you want to talk about that?

Denert: I think we need a good understanding of a lot of theoretically based methods. And a very pragmatic approach of applying them to real-life problems. I don't believe in a very strong formal application of methods and languages. You must be very pragmatic at least for the next ten or twenty years—perhaps forever. But we have to have a repertoire of such things, and apply them in a creative fashion. That means we need very good, very well educated, trained people; and it doesn't work the way that a lot of managers would like to have. I have those people and I have to live with them. The methods and tools have adapted to their low skill. Making software needs high skill. With bad people it doesn't work. So I have an attitude of an elite.

Aspray: But I suppose that for the large applications company that has a small programming unit, it is hard to attract quality people.

Denert: Yes, that is their problem. But that doesn't change the situation. It doesn't change my statement that it's very complex and very difficult to build big complex systems. Part of the solution would be to buy standard software. Not to develop software, but to buy it. We have a very successful company in Germany, SAP, which is very good. So it would be a solution for a lot of smaller companies not to develop software at all.

Aspray: How do you view the software development process? Where does it stand in comparison to these other enterprises?

Denert: It is not comparable with production in the factory. There are some people, especially in Japan, who talk about the software factory. They tried to say that software has to be produced like mass production of cars. That's nonsense. We don't have mass production, we have the same process as it happens with developing new automobiles. There are laboratories and there are development departments and divisions. They are building prototypes and testing them. It is also a very creative design and engineering task.

Aspray: To put it another way, software production, software development, in your case is the equivalent of design?

Denert: Right.

Aspray: In mass manufacturing complexes?

Denert: Right. What we don't have is the mass manufacturing behind that. Companies like Microsoft are providing you with books and some diskettes. They have to produce them, but that's not very important.

Aspray: That's very routine. Maybe the closer analogy to mass production and the software field is with companies that believe that you can build discrete subunits of program code which can be recycled over and over and over again. Then you just put the modules together in different ways for one customer after another.

Denert: We have not reached the state where you can do that. That's an idea connected with object orientation. Cox talks about software ICs, software integrated circuits. Putting objects or classes together like integrated circuits on a board. We have not reached that state.

Aspray: I know one of the American companies has been touting their ability to do that, SofTech. But I don't know enough about the company to know to what degree they achieve it.

Denert: They may very well. For me they were better known in the 1970s. They had good advertising on the backside of *Communications of the ACM*. Doug Ross was very well known. But in the meantime I lost track of them.

Aspray: I don't know either. I know Doug. But I don't know about this matter. Another thing that struck me when I was looking through your literature was the visual images you use in your advertisements. The pictures of the different aspects, features, qualities that software should have.

Denert: Yes. Six nice pictures. [chuckling]

Aspray: Are they tied in any way to a philosophical understanding of what software is all about?

Denert: I don't know if you have read these texts. They are very good German texts. One text is about the tree, and the corresponding one is about software. That is the pattern throughout. It is an image. We try to say that we like nature and that there is some relationship between software and nature. Good principles of nature can be applied to software systems. But that's very philosophical, something substantial to apply in our daily work.

Aspray: Are these concepts yours?

Denert: Yes. We developed them. There was no public relations officer. Maiborn did this series. I made another series earlier. Maybe it's interesting for you because it had something to do with history. [laughter]

Aspray: I'd like to know about that.

Denert: There was a man who developed these small texts, which are very good. We have somebody who makes our corporate identity and prepares brochures. He brought the pictures and made the whole thing. We did it ourselves.

Aspray: Are there some other comments you want to make?

Denert: I hate CASE tools because of their unsatisfactory methodology, like structured analysis, which is wrong in my opinion. As far as one can say, the methodology is wrong. It's not mathematically wrong, but it's not a good methodology. A good methodology is based on the ideas which were established by Simula, Parnas, Dijkstra, and others. The idea of object orientation is the main thing. The methodology of structured analysis was described by Tom De Marco in 1978 in a very famous book, but nobody used that methodology until it was implemented in CASE tools on PCs. It was not chosen because it is such a good methodology, but instead because it was possible at the beginning of the 1980s using PCs with graphic interfaces to build tools for drawing software. I told you I made my Ph.D. dissertation on a two-dimensional programming language with which one could draw programs. It was a bad idea. They retried that idea. Drawing, drawing, drawing. Entity relationship diagrams. That is quite reasonable, but the bulk of data flow diagrams are a terrible idea. Nobody would consider using that methodology with pencil and paper.

But suddenly they had PCs and they could draw small pictures on it. There is a new book by James Martin, who influenced the CASE market very much because he is somehow the father of ADW, one of the very successful CASE tools. And IEF, marketed by Texas Instruments. And also Excellerator. He has influenced this field very much. In his new book, he has a chapter entitled Killer Technologies or something like that. These are things he believes are important. There he states that "visual programming" is very important. This is the idea that one draws software. He even says that future software engineers will use sound for understanding their software design. I think he's crazy in this respect. I believe that it doesn't work. The bulk of the work in software engineering has to be done by writing structured texts. Some sort of specifications. Programs are finally texts. The main work has to be done in some textual form. You have to have good pictures, good illustrations of these things, but I do not believe that the main thing is drawing pictures. I have a lot of pictures in my book. Everywhere there are pictures, graphical illustrations. They are very important, but they . . .

Aspray: But they don't replace the real texts.

Denert: Right, I believe that and I know that I am a little bit against the mainstream. But perhaps the mainstream is changing its direction.

Chapter 3

Kurt Schips

About Robert Bosch GmbH

Founded in 1886 in Stuttgart by self-educated electrical engineer Robert Bosch, Robert Bosch GmbH has been a recognized leader in the automotive components industry for a century. Its first innovation in the automotive field was a hand-crank motor starter introduced in the early 1890s. Within two decades, Bosch had become a world leader in ignition systems.

Bosch survived desperate economic conditions and international embargoes against Germany during the two world wars because of its penchant for innovation and its commitment to producing a diverse, high-quality product line. Strong growth in the 1930s prompted corporate reorganization into a private limited company in 1937. Within this corporate atmosphere, Bosch made numerous automotive innovations, including the first injection pump for high-speed diesel engines, electronically controlled gasoline injection, and antiskid braking systems (ABS).

In the 1960s Bosch increased the resources it devoted to research and development (R&D) in the electronics field. Since then it has entered the markets for mobile cellular telephones, car stereo systems, and a variety of telecommunications products. Bosch restructured its telecommunications activities in 1989 to form Bosch Telecom, a worldwide supplier of telecommunications equipment operated by over forty thousand employees.

Today Bosch is an active market participant in electronic and mechanical automotive equipment, communications technology, power tools, household appliances, thermal technology, plastic products, packaging machinery, industrial equipment, hydraulics, and pneumatics. By 1992 Bosch employed over 180,000 people in 130 countries. It has subsidiaries and foreign interests on all five continents and seventy factories around the world.

Kurt Schips

Place: Gerlingen, Germany

Date: July 5, 1993

Schips: You know why I was smiling when you mentioned Japan, the United States, and Europe as you described your project? The reason is that, a couple of years ago, we discussed this issue in our company of why is it the Japanese are so successful? That discussion went on and on and on. Some of my colleagues thought they knew everything about why the Japanese are so successful. After some time I really got fed up. I thought I should spoil the whole affair a little bit and said, "Now I know exactly why the Japanese are so successful. The reason is very simple. I personally met most of the leading managers of the electronics industry in Japan. What they all have in common is that they are elderly people and have basically no knowledge of English at all. Therefore they never get information about modern management. They were running their companies like an old battleship, that is why they are so extremely successful."

I mention this only to spoil a serious discussion. But after all, I am not so sure if there is not some truth in it. I think one of the reasons in Japan—as compared with Europe and especially as compared with the United States—is that they are not influenced so strongly by the financial results in a short-term way.

They deliberately or unconsciously are looking for a market share and then they transfer it into a profitable business. That's one explanation. That is why I smiled when you mentioned this.

Aspray: Do you think that's true also of German companies? After all, between the end of World War II and now they built up from a devastated economy and became very powerful.

Schips: When I compare Japan, Germany, and the United States, I see Germany in the middle. Not extreme financial treatment, nor extreme technology treatment, it is somewhere in between. The success the German industry had after the war was based on several factors. They had good human resources and they had a chance to start at the beginning. They had to buy new machinery, so equipment was relatively new. The people were highly motivated to survive. Many of our problems today are coming from rules that allow people to have enough money even if they are without any work. Even then, they can still make their living. That's not a very strong impetus for the population as a whole to be as creative or as active as in the years after the war.

Aspray: Can you tell me about your own career? Can you tell me when you were born, where you were educated, what you career path was?

Schips: I was born here in Stuttgart in 1927. I grew up in Stuttgart and received my high school education here. At sixteen years of age I was called into what they called "the labor service" and later on the army. I was an American prisoner of war. I was discharged in the summer of 1945. I opened a radio repair shop at age eighteen in one of the smaller cities in the environment of Stuttgart. After a year or so I had the chance to go back to high school again. I finished high school and started to study electronics in the Technical University of Stuttgart. I finished in 1952. At that time we had a recession. My father was working for Bosch for forty years, therefore I asked for employment at Bosch. I asked for a job in technical sales in exports, but at that time there was not much of an export business. When I was invited, I was told, "Well, we have only one vacancy in our company, which is in the patent department." I looked hurt, but had no choice but to say yes, and started to be trained as a patent attorney for four years in the patent department.

I started to like this profession. After four years I had the chance to become head of the patent department of one of the subsidiaries of Bosch in Hildesheim, named Blaupunkt. I served there for eight years. At the end I was responsible for patents, licensing, and advanced technology research. That was my first managerial experience. Besides my activities in the company, I studied economics at the University of Göttingen.

Aspray: Why did you do that?

Schips: As an engineer I did not like to hear my law-trained and economics-trained colleagues talking about things I did not understand very well. So I said to myself, "Let's try and see if I can adopt this knowledge too." I did it in Göttingen. But before I could gain my master's degree in economics I got the opportunity to become a technical manager of a subsidiary of Bosch in Berlin. That was in 1964. It was in the times just after the erection of the Wall. It was a somewhat chaotic situation, but I liked it. It was far away from headquarters. I had relatively great freedom to develop the company.

But in 1968 I was asked to come to the headquarters as head of all patent and license operations of the Bosch Corporation worldwide. I took over more or less reluctantly because it was narrowing down my freedom and the scope of my activity. But I was asked to do it and I did it. After two or three years I took over, from one of the members of our management board, the responsibility to supervise one of the smaller subsidies of Bosch in the field of electronics and what we call investment goods: packing machinery, industrial equipment, and so on.

Gradually, I took over more and more managerial responsibility. At the beginning of the 1970s I was appointed as an associate member of the board of management. Step by step I was promoted to vice president, executive vice president, and the last position was called senior executive vice president. In my last position I was responsible for all activities of Bosch in the field of communications. There are about forty thousand employees and a 6.5 billion Deutsche Mark business. Besides that I still had the responsibility for the patent and license activities of Bosch. I also had to look over all construction activities of Bosch, including investment in land and buildings as well as what you may call "investment controlling." You know, we are highly diversified and it is always a problem for the central management to allocate the money to the various divisions. Part of this task was to assign investment funds to the various activities.

Aspray: Let us talk about patent strategy within the company. In a field such as electronics, where things are changing rapidly, patents themselves may play a different role than they do in more stable industries. Can you talk about the way that patents operate in your business?

Schips: In this company, which is now more than one hundred years old, it is remarkable that patents played a role from the beginning. The founder of the company, when he decided to go into automotive electrics, established patent research before he became active, which was quite unusual in those times. Already by 1910 the first employee was in the company doing patent affairs. It developed gradually. One of my first tasks was to formulate the strategy of patent and license work within the Bosch group. We decided that the first priority of our patent work was to keep the scope of our activities free of hindering patents. The first thing is either to destroy patent applications that were relevant to us or to buy rights on it.

The second priority was to safeguard the results of R&D within our company by application of patents in Germany and so on. And the third priority was to make money out of our patents and our know-how. We established a clear roll of priorities, which are now common in the company.

Aspray: Perhaps you can say something about the organization and management of advanced research and development in the company. Is it centralized?

Schips: We have in many ways a somewhat unique structure in our company. The basic units are the divisions. They are responsible for R&D, production, and sales. When they are big enough—and most of them have the proper size—they fully have to look after their own R&D. But in the case, say, of the group of communications-orientated divisions there is a need to have some overlapping R&D activities—things which do not fit exactly in division one, two, or three. This is done by what we call an Advanced Development Department. That is the second level. And for all divisions we have a central research unit. This central research is mainly maintained to look after basic technology and materials, which are so expensive that you cannot handle them in an advanced development department, or in the R&D departments of the divisions. So we have three layers. But that is not a hierarchical structure. Most of the decisions are made in the divisions because our belief is that the divisions should act like independent companies in order to be flexible enough to cope with the demand of the market.

Aspray: If one division's researchers come up with some promising ideas, how are they communicated to other divisions at the company?

Schips: That is the task either of the advanced development departments of a group of divisions or of central research. They have to watch what is going on. They should know it very well. When something comes up that is important to other divisions, or to other groups of divisions, they will learn about it and then they have to talk amongst themselves to find a solution. If there is no solution, the problem was to be solved by central management. But usually we can rely upon the common sense of our managers.

Aspray: What is the process like of taking a promising idea developed by the researchers to product development?

Schips: That's a twofold process. From time to time, central research presents their results to the development department of the divisions. Then a division can ask to get the results. That's one way. The other is they have to do some sales work, when there is not an obvious demand. As you can imagine, this is not always successful. Our research people have ideas, they come to some results, but sometimes the results are not very appealing to our people on the front of the business. I think that's a problem we all have in organizing R&D.

Aspray: Do the researchers travel to the next stage with a project typically? Or do they pass it on?

Schips: Sometimes they pass it on. There is a desire of the central management that the people of the central R&D or of the advanced development departments go with their results into the divisions. Otherwise, the transfer of know-how is not efficient enough. On top of this, we have to take care that employees in Central Research are not getting too old. Therefore we need fluctuation. Here we have to work against human behavior to stay where they are, where they are successful and rewarded. It is the task of the central management to stimulate fluctuation between various layers of R&D.

Aspray: At what point is so-called manufacturing engineering brought into play?

Schips: A second unit of about the same size as Central Research is Production Technology. That work is done only on a central basis. They develop new soldering technology, welding technologies, stamping, and pressing processes. These guys at this central production technology development organize at least once a year meetings of all what we call "production planners" at the division level. They join them, present their results, and give them information about recent developments within the company and outside. They also have to serve as lookout for activities outside of the company—to watch the general technology progress. They tell you their findings in these meetings. That's one thing. The second is that every division, once a year, gets a visit from some of the central people. I would not say it is exactly an audit, but it comes close to it.

In central management we can ask our people at the central level how production technology is developed in a certain division. So we know fairly well how far advanced the various activities are.

Aspray: The actual design work of taking the idea of a product that has been approved for production, but for which there is still a lot work to figure out how to make it properly and get quality, does that takes place on the division level?

Schips: Only on the division level because they have the link to the market. It never will take place at the advanced development or at the central research level.

Aspray: I don't know the Bosch company very well, but my impression from what you are telling me is that the company's philosophy is to keep the R&D in-house, not to simply be an acquirer of technology. Is that a correct assumption?

Schips: That is correct, but we also try to acquire technology from outside. This is not self-sustaining because of the not-invented-here attitude. The management has to push the engineers from time to time to go outside and look for what is already available before they start an expensive R&D project. When it comes to a product, it must be done in-house because otherwise we cannot guarantee the quality standards we are looking for. The typical design of the product must be done within the company.

Aspray: In what kinds of circumstances would you decide to acquire licenses from some other company or even purchase a small company?

Schips: When we see there is development that could be of interest for us or if there is a patent which may hinder us in the future, then we approach the owner and start license negotiations or even take-over activities—but not in a hostile way. We avoid that and we are successful doing so. We always try to come to a mutual agreement.

Aspray: What do you do in order to enter certain kinds of markets in different geographical regions? Or specialized industries where you have to know something about the industry? Is that a case where you commonly enter into agreements with other companies?

Schips: Yes. For instance, when we approached the United States market for the third time. The first time was 1910, and we lost everything during World War I. We started again in 1920, and we lost our property in 1945. We had to find our place in the market again, and we ran across other companies which developed when we were not active in the United States. For instance, one of our main competitors had developed a principle for electronic injection and had a strong patent position. So we had to negotiate with them and come to an agreement.

Aspray: What is the product life of most of the Bosch products? I am asking because of some questions I want to ask about the research process.

Schips: Bosch is a highly diversified company, as you might be aware. 50 percent is automotive equipment. Roughly 25 percent is communications. A further 25 percent is household goods and investment goods. Depending on the product line, the life span of a product is different. One extreme is consumer goods. For instance, in the field of entertainment—electronics, television receivers, car radios—the life span for a product is typically less than two years. But in the investment products field, the life span can be fifteen years. Packing machinery could be made in a similar shape for a decade or even more. With our automotive products, I would say the average is five years.

Aspray: For those parts of your product line that have short-term lives, what effect does this have on your R&D process? You must have several projects along the way at different stages. How do you manage that?

Schips: This is again done by the divisions. The management of the division have to look after their own R&D work. There are some projects close to implementation in the market. Others are in a more or less premature stage. Usually in any division management, we have a technical manager. In the larger operations we may even have two people: one looking after R&D and one for production. These two people, or one person in the smaller units, have to look after these R&D activities and are responsible that at the proper time the necessary products are developed.

Aspray: Is there an accepted amount of time when the director of research in one of these divisions is talking to his research staff about project completion? Is there an expectation that this research will have a payoff in five years or three years or ten years? Are there rules about that?

Schips: When we start a project, the aims are specified from a technical point of view. But also from a sales point of view: what turnover can be expected, at what time the product should be available for the market. At the very beginning of the project they define these aims. Unfortunately, there are sometimes delays—unexpected problems or additional wishes coming from the sales people. There is a steady struggle between the people involved. One of the basic problems in modern industry is to be at a very early stage on the market.

Here in Germany we have a special problem which you do not encounter in the States nor in Japan. You may have heard that our unions are very strong in favor of the so-called 35-hour week. That means that the average German employee is working every year about 1,600 hours. In Japan they still work 2,100 hours. Most of the people are looking into this problem from a cost point of view. But I think that's only half of the story. Maybe even less than half of the story.

In my opinion, the greater effect of these shorter working hours is that development departments cannot be so fast as development departments in Japan. When you start from the assumption that the quality of the engineers in Japan and in Germany are at the same level, which you can dispute but I think it's at least initially a fair assumption, then the development engineer in Germany is available at his working place only 1,600 hours a year. But the development engineer in Japan is available 2,100 hours. Many of our development engineers are on a voluntary basis working longer hours, but that applies as well for the Japanese. The gap you figure is 500 hours that the Japanese are working more. In these extra hours, they can speed up development. And in the next year this comes on top of it. After a couple of years, you find that the Japanese are much faster in the market than the Germans. That is a problem I already see now, but in the future it will even be more serious.

Aspray: Why has Bosch made the decision to be organized as a decentralized company?

Schips: For flexibility. We felt that at a certain point—and it was at the initiative of our chairman—we were reacting too slowly to the demands of the market and therefore, the idea came up. We created divisions.

Aspray: What aspects are centralized within the company? Do you centralize capital, do you centralize any marketing or other functions?

Schips: We centralized legal affairs, patent operations, and financial management. Every division has its own head of economics. But he is on a very short lead. If there is an excess of money in a division, this money goes to the central financial department. We also have some centralized marketing, especially if the name Bosch is involved. There should be rules to be followed in any division in order to have a unified appearance to the market. Everything else we gave to the divisions.

Aspray: I know that some electronics systems manufacturers are now also starting to centralize purchasing because so much of their buying is semiconductors, which make up a lot of the cost of the products.

Schips: Quite right. We have a central purchasing office which relies on the purchasing offices of the divisions. But when it comes to, for instance, steel, copper, or semiconductors, they coordinate the division level of purchasing offices in order to use the benefits of buying power.

Aspray: You suggested that there was another way that the company was decentralized in terms of its product lines. Do you want to speak to that?

Schips: That brings us back to Robert Bosch, the founder of the company. One day he felt he was too dependent on automotive products. Automotive was not only a success story. There are some ups and downs in between. Therefore he started to look for other business. That was the origin of activities of Bosch in the field of entertainment goods. There was also hydraulics (not only for automotive products, but also for machinery), manufacturing equipment, and household goods. With a smile I would say unfortunately we were not successful enough because the automotive side of our business grew also. Sometimes extremely fast so we never got more than 50 percent of our business to be nonautomotive. At the moment we have about

50-50 between automotive and nonautomotive. It remains to be seen what the future holds.

Aspray: One can live with that kind of success.

Schips: We can, of course. We always try to do more, but the automotive side was successful and others could not overcome them.

Aspray: In choosing new business areas, does there have to be some sort of relationship to old businesses? Some synergy between them? Some sense of knowing how to manage them? What considerations go into those?

Schips: We always try to stay within the scope of technology known to us. We have a German saying, which is hard to translate. It means "Shoemaker, continue to work on shoes; do not make fancy things." That is a guide for our selection of new products. Hydraulics, which I already mentioned, is typical. We used this technology for automotive equipment and now we use it also for other machinery. Another case is the refrigerator. The basic elements of the refrigerator are the motor and compressor. For both elements we had the technology to produce. Another example was optics, used for headlights. You can also use optics for other products, so we started a photo business. We had a fairly large operation in the field of hand-held movie cameras, employing up to four thousand people. That came out of our optics know-how. In a movie camera there are moving parts. You have a similar technology in packing machinery. So another new field developed. But sometimes we had marketing problems when we ran across different behavior in markets new to us.

Of course we were not always successful. We once started a business of prefabricated housing. Although it was somehow related with something we already did, it was a failure. Here we were too far away from our existing business and from the experience we had.

Aspray: Has the company left business areas in your time?

Schips: Yes, that is natural. When you look at a tree it is growing, but it is also giving away, not only leaves, even branches. So that is what we do. We acquired new branches, new activities, but at the same time we gave up—stopped or sold—business. We started business in the photographic movie area. We did it for sixty years or so. And after the photographic market decreased, especially for movie cameras, we stepped out of this business. So we feel strong enough, even after sixty years of activity in a certain market area, to step out. We think that is vital for every company that likes to survive. You cannot stay on your old original product and only develop it. From time to time, you have also to cut things. You cannot only get new products.

Aspray: What kinds of decisions have to be made in allocating capital to your various divisions—as seen from the top side?

Schips: Every year, every division has to make a proposal for the budget of the next year. Included is also a view for the next three years. So every year central management gets a plan from every division: what they expect to do, what turnover they expect, and so on. Then this plan is discussed—this happens every fall—between the central management and the management of the various divisions. After this process, these plans are united in a plan for the whole company. During this process the central management has to make a decision: in which division or which product line is it useful to invest money and to which amount?

Aspray: I know it's an important activity, but how routine is it? Is it like creating the business anew each year?

Schips: It is for the division managers the most important thing every year. They should feel like an owner of their business. They have to defend it to the central office. So there's a struggle going on. That is not done in an easy way. Certainly the central management must be flexible enough to make adjustments each year.

Aspray: In some sense a lot of your budget is already set because you have made commitments in the long term to capitalize or with product development and so on.

Schips: Yes.

Aspray: So there's a fair amount of stability there.

Schips: That's what we try to achieve: stability, because we have a preview to the following years. But we also have flexibility because every year is on the test bench, must be discussed—even if in the year before it was decided, "Well, there is a five-year program."

Aspray: On the Bosch board of management, what is the background of the people that hold those positions?

Schips: We have people trained in economics, law, marketing, and engineering. All sorts of people. What may be a little bit unique is that some of them have line as well as staff responsibilities. In my case I had line responsibility for all the communication subsidiaries of Bosch, while my staff function was patent and licensing, investment, controlling, and so on. Some board members have only staff functions and some only line functions. But most of them have both. The thought behind that approach is that when you are responsible for both line and staff, you are not so much inclined to look into things only from one point of view. That adds some balance in thinking, which is implemented by this structure.

Aspray: I see. What kind of technical knowledge do people have to have to be members of this board?

Schips: Today, all of them are trained in a university. They all have doctorate or master's degrees from a technical university. In the past, we called them engineering schools. You do have not a similar institution in the States. They made a career by first being an apprentice and then going to school. This was a school something between high school and university. It was sometimes called "academy" in Germany. They were promoted as "engineers" without diploma. So they came from the work bench and have been successful managers in our company starting as apprentices. For instance, our manager who looks after marketing in general started as an apprentice of the company. Later on, he went to the university. This is not unusual. We try to remain open for everybody, but for the time being, most now are coming from the university.

Aspray: In doing your own job, what difference does it make that you have an engineering background?

Schips: All this management education brought up the idea—at least in Germany but I think also elsewhere—that a good manager can run every company. I think that's a fatal mistake. That is true when he competes against another "general" manager. But when he has a competitor who has managerial experience and specific experience in his field, in his product range, then he is always going to come out second. It is a mistake which occurs more and more frequently, unfortunately. People believe they should look for a good manager, who will get the wisdom necessary to do the job, from one day selling mineral water and the next day running a company with electronic products. Again, there are examples which prove that it is possible. But in general, it takes so long—even when he is excellent—until he gets all the information he needs. In this time the company has a high-paid apprenticeship at the top.

Now I'm coming back to your question. I was responsible for the communication activities of Bosch. At university I learned the technologies in this field. So when I had discussions with my division managers, I was always regarded as somebody who at least had a basic knowledge. I think it depends on the company, but in a technology-driven company it is necessary at some place in the top hierarchy to have technology background.

Aspray: But not everybody on the management board needs to have it.

Schips: No, not everybody. Certainly not.

Aspray: Assuming that your personal view that you just expressed is a company-shared view, top management-shared view, what does this mean about the appointment of people to higher-level positions in the company. Are they done from within primarily? Or do you bring people from the outside?

Schips: It is done primarily from inside. We do not exclude people from outside. I would guess 10 percent, maximum 20 percent of the managers—probably closer to 10 percent than to 20 percent—come from outside.

Aspray: What does the company do to prepare engineers to move up through the ranks to be managers?

Schips: First, the most important thing is training on the job. When an engineer comes from the university and starts his career, let's say in development, he has to learn about the dos and don'ts within the company, company structure, and so on. That's extremely valuable. I mentioned I went to university at Göttingen to study economics. It was impressive for me to see how much I have learned about in practice before I was exposed to it from an academic point of view. Young engineers learn a lot during the first years when they are open-minded.

When they show some potential to become a manager and show an interest to do this, then we have several promotion programs. We send them to attend management programs in Europe and the United States. MIT, for example. I once attended a summer course at Stanford. Also in Europe, Fountainbleau for example. These are preparations for people who are on the promoting list. We send people away, but we also have a Bosch College, as we call it. That is a very useful tool. It was drafted especially for the purpose of upgrading people who, let's say, have ten years or fifteen in the company, have made their ways through the ranks, but have lost connection to developments which were going on in universities. For instance, computer knowledge, or organizational methods, or basic mathematics to refresh the knowledge they once learned at university. The training at this Bosch College is run by an internal group of people, but the teachers are coming from universities. We try to make our managers capable to cope with the problems they come across in their daily work.

Aspray: I don't want to prolong this discussion too long because I know you have other things to do. But I have two more questions for you. The first one is: what is your greatest challenge in doing your work on a day-to-day basis? What's the hardest, most subtle kinds of issues that you have to address?

Schips: I think for a manager in general the most rewarding and the most important thing is to safeguard the future of the unit he is responsible for. Including all the problems involved. From people, to financial resources, to pushing development, to looking after lean production.

Aspray: My last question brings us back to where we started which is, do you want to say anything more about your competition from Japan or other parts of Europe or the United States?

Schips: Let me start on this question with two remarks. When I started my career here at this company, our most important competitor in the automotive field was a company in Great Britain. We had a high regard for them, and we were even working together in certain fields. They were good competitors. Reasonable ones and from a technology point of view, very advanced. What happened to them? They are almost out of business in the meantime. Why? Not mainly because of "British disease." Mainly because their main customers disappeared. The automotive industry in Britain fell down, although not completely. Having relied mainly on their local customers, they still have some market share, but they are no longer the dominating competitor. That is one observation.

The second is, when I came to Blaupunkt, an entertainment goods company, and I went to a trade fair we had at least thirty or forty competitors in Germany. When I go to a fair nowadays, there are only two or three competitors. And one or two of them are foreign. You see the big change that is occurring. We have to live with it. We are not, by far not, at the end of the period we are in now. After this recession we face here in Germany the economic and the industrial landscape will be entirely different.

In worldwide competition I think Germany does have a good chance. But we must consider that the Japanese are extremely strong. They learned their lesson. When I first came to Japan thirty years ago, I was a lonely scout. When I told my colleagues that there is something building up, they nodded, took note of it, but it was not something that was of high emotional value. That changed. From an industrial point of view they are well organized and extremely fast. Fast is the key.

The United States industry, which was, after the last war, the big example to be followed for all Europeans, advanced production technology and advanced managerial knowledge. Like pickers, we went to the United States and tried to learn it. But American industry fell down a decade ago. There was an extreme weakening of industry in the United States in general. But we notice it is improving again. How far it will improve remains to be seen, but American industry is definitely improving its efficiency. Maybe not only the efficiency. They lost the long-term perspectives in their environment. The take-over battles, according to my personal opinion, were absolutely crazy. It was a waste of energy and financial means which you can hardly understand. But American industry is coming back. The European industry is, again, somewhere between the Japanese and the Americans. What will be seen in the next years is hard to predict. If the free trade formula will succeed—German industry is very much in favor, but German industry is not Europe, we are part of Europe and we have to accept what the majority is thinking. There could be a change. It could also come from the United States. You still have your protectionist approach, very strong. Some of your senators and trade representatives still are always thinking of how they can protect farmers and parts of industry. The Japanese are doing it their way. Maybe it is not what sometimes was suspected, that there is a master plan in Japan. But they are used to living together, and have done so over many centuries, and all of the Japanese believe, "What I do not seed in the spring I cannot harvest in the fall." It is experience that we forget sometimes. They have a vision. At least in Europe, we have a lack of vision. Maybe you Americans are more open-minded. You can be more easily motivated, in contrast to the highly sophisticated thinking which characterizes Europe. Maybe that is an advantage in the future. Life shows us that there's always a new way of advancing.

Chapter 4

Arno Treptow

About AEG

AEG began in the 1860s when founders Emil Rathenau and Julius Valentin purchased the engineering factory of Webers, a small producer of engineering parts for ironworks. After a profitable wartime period in the early 1870s, the company went public as the Berliner Union. The brief wartime boom quickly turned into a nationwide economic crisis, forcing the Berliner Union to fold soon after its establishment.

Disillusioned with mechanical engineering, Rathenau turned his attention to the exciting new phenomenon of electricity. In 1882 he formed Studiengesellschaft, a consortium armed with the Edison patent licenses for France. Studiengesellschaft obtained the license to provide the street lighting for most of Germany. In 1883 Rathenau established Deutsche Edison-Gesellschaft, gaining the sole right to use all Edison patents in Germany. First installing lights in factories and steamships, the company turned to the construction of central power generating plants and built Berlin's first central generating station.

The opportunity to expand worldwide came as DEG and Compagnie Continentale signed a May 1887 treaty ending their formal partnership. DEG changed its name to AEG (Allgemeine Elektricitats-Gesellschaft) and initiated a rapid expansion plan. Through internal growth and acquisitions, AEG grew steadily until the Second World War. It worked in many areas of electrotechnology, including theatrical lighting, the design and construction of power generating stations, light bulb and arc lamp manufacturing, and electric trains and tramways in the late nineteenth and early twentieth centuries.

During the Second World War, AEG lost nearly all of its production facilities in Germany, Poland, and the Soviet Union. The company tried to expedite its own recovery after the war by purchasing more than fifty firms in the household goods industry. This strategy initially proved beneficial as an improving German economy put more spending power in the hands of German consumers. The recession of the 1970s and heavy involvement in the construction of nuclear power stations proved disastrous, however, abruptly ending AEG's recovery.

Despite several efforts throughout the twentieth century to maintain its corporate independence, it became increasingly clear that survival depended on participation in joint-venture schemes. AEG entered into a telecommunications deal with Bosch and video contracts with JVC, Thorn-EMI, and Thomson-Brandt. Nonetheless, AEG was unable to remain solvent in tough economic times. It became one of the first of many German economic casualties in the 1980s. AEG barely survived bankruptcy proceedings, having sold off nearly all of its assets. A new era for AEG began the following year as Daimler Benz purchased 56 percent of the company, wiping out all of its previous debts.

Currently, AEG maintains controlling interest in one hundred companies in 107 countries around the world. It promotes and maintains a direct export business from Germany selling systems for power transmission and use, automation, industrial and environmental protection equipment, rail systems and components. Its foreign subsidiaries engage in a broad spectrum of independent operations ranging from engineering and delivery of complete plant, systems, and components to subcontracting for international projects. AEG is known worldwide for its electrotechnical systems and components, domestic appliances, microelectronics, and rail systems, and by the end of 1992 employed nearly sixty-one thousand people worldwide. The company prides itself on its continuing commitment to Emil Rathenau's desire that AEG remain a company whose primary interest is electrotechnology.

Arno Treptow

Place: Frankfurt, Germany

Date: July 1, 1993

Aspray: Could we begin by having you tell me about your education and career?

Treptow: I was born in 1930 in a city called Kohlberg, which is just beside the Baltic Sea in Pomerania and which now belongs to Poland. At the end of the war, I was a boy of fourteen. I went to a small village in the northwest of Germany, where I finished at the Gymnasium with what we call the *Abitur*. At that time we had to gain half a year experience working in a company. So after half a year of work I started studying electrical engineering at the Technical University of Darmstadt. I had some health problems there. I spent two years in the hospital, but it was not a major threat.

Before that I was lucky to receive a scholarship, *Studienstiftung des deutschen Volkes,* because I was a fairly good student. I finished my university study in 1958. For my diploma, I worked for four months at AEG in connection with the university. They had a problem and went to the university for help, and I was assigned to this problem. So I worked at AEG and solved that problem. At the end of my four months they gave me a very good offer of employment and asked me to stay. And I did so.

Aspray: Was this work on the power side or in telecommunications?

Treptow: It was on the power side. It was a special problem concerning DC high-speed circuit breakers.

Aspray: So then you came to work as a regular employee?

Treptow: Yes. I worked at AEG in the research and development department for several years. At that time I got in contact with the International Electrotechnical Commission (IEC) in Genf. This was part of my job. After some years in research and development, I changed over to manufacturing.

Aspray: Before we move on to that, what were the kinds of problems you worked on in research and development?

Treptow: It was more development than research. It was development of circuit breakers, or more specifically, addressing problems with the arc inside the breaker. It was mechanical and physical.

Aspray: Was there much that you would consider research rather than development going on within AEG at the time?

Treptow: No. At that time in that division, we were a crew of several young engineers. We did a real push in technology at that time. We got a lot of patents and we were creative. It was a nice time. After that I changed to the manufacturing side.

Aspray: Why did you do that?

Treptow: I did it because in that development division my boss was only two years older than me. The general manager asked me if I wanted to change to another sector. He offered me the job in manufacturing. It was just at that time that we finished the development, more or less, for those products. In order to bring them to manufacturing I thought it was a good situation to go with the product and try to manufacture it.

Aspray: To see it through.

Treptow: I did this for three or four years. One year later I was already the boss for manufacturing for this plant. Three years later I became the general manager of the whole plant.

Aspray: You moved through very rapidly.

Treptow: Yes, I did. I was forty-two years old. I was the general manager of that division. It was a good job there. I liked it. Three years later I was promoted and sent here to Frankfurt as part of the management board. At that time, it was 1976 and I was forty-five years old. In 1976 AEG was just restructuring. The whole company was divided into four companies. In one of those companies I was a member of the board, with responsibility for manufacturing. Then I became a general manager for one sector of AEG.

Aspray: Is that where you are today?

Treptow: No. In 1985 I joined the board of management of the whole company.

Aspray: What are your main responsibilities today?

Treptow: Today in my group there are two divisions. One is called power distribution and is concerned with high-voltage circuit breakers, high-voltage network and systems, and transformers. The other is our component business. That is what we call the low-voltage business. Circuit breakers, conductors, motors, drives, and lighting system meters. On this side I was leading a group of specialists concerned with manufacturing systems.

Aspray: Manufacturing is a topic that is on everybody's minds these days. Can you tell me something about the kinds of issues that are important to AEG in this area? What kinds of issues were you looking at in this group? What was your strategy?

Treptow: There is a general view, but at the end it is based on the type of product one is dealing with. For instance, in power distribution, that is an international business. We have here in Germany the main center of research and development in this field. We manufacture here what we call the noble parts. But we have small activities in several countries around the world. We send the noble parts to them, and they make the adjustment to the customer's requirements. This is a special business because there are, more or less, only eight companies in the world in this business. That's not quite correct because that is without Russia, the old Eastern bloc. But these eight suppliers (five are European and three are Japanese manufacturers) reach a very high technical standard. There are currently no Americans here in the business. It is more or less a Japanese or European technique.

Aspray: The products are made available through cross licensing?

Treptow: That is correct.

Aspray: How do you differentiate yourself from your competitors, if it is not mainly on technological factors?

Treptow: It depends on the relationships between suppliers and customers and also on technological points. It is not a field which changes every year like the personal computer business. This is a business involving products with much longer lifetimes. And there is a reason for it. Maybe you remember, some years ago there was a blackout in New York. The utilities are very careful when they introduce or change the technology every year, and they are very keen to have the same supplies and ability to minimize any big change, as one finds in some other parts of the industry.

Aspray: I would think that in a situation where there is not rapid product obsolescence that there is even more consideration of things like manufacturing and keeping costs down and keeping reliable products.

Treptow: Yes, it is one of the most important aspects in the business. The other main aspect is the supplier-customer relationship. This is more important than economy of scale. For instance, if the Japanese people come here and offer a circuit breaker that costs 20 percent less, the utilities wouldn't buy it.

Aspray: Because you have such a strong working relationship with them?

Treptow: Yes. And they already have a certain kind of equipment installed in the field. So why should they change it? Then they would need two stocks for repair material. The service would be much more complicated.

Aspray: Do you actually have some of your own employees on site at your bigger customers?

Treptow: Sure. Everywhere.

Aspray: Could you describe some of the landmark changes you have observed in your field?

Treptow: This is one field where we have done a lot of research. In the past we have seen many changes in technology. Twenty years ago there came a new generation of circuit breakers. It was the SF6 technology, which was the change from air to gas inside the circuit breaker to distinguish the arc. This was a big step in technology. Then maybe ten or fifteen years ago they started to use this gas for insulation in power stations and distribution centers, which makes it possible to build the equipment much smaller. So there was another big step. Development continues to make it incrementally better and better. It also promotes better services and handling. It's not the basic technology, it's more or less the incremental improvements.

Aspray: How far in advance can you see a major change?

Treptow: Presently in this field of circuit breaker technology we have two technologies, SF6 technology and vacuum circuit breakers. At the moment they are competitors. Some customers prefer one and some prefer the other. I do not foresee a big change in technology in the next five or ten years.

Aspray: Do you have people in your laboratories looking for that next step, though? Working ten or fifteen years ahead?

Treptow: Sure. As you know we are part of the Daimler Benz group, where all research labs are combined. But inside this group, we have research people who work more or less for the AEG side.

Aspray: Is their research done in different locations?

Treptow: Yes.

Aspray: Are there relations between the people in the advanced research area and the field people?

Treptow: If people in the field have problems, they talk to the research people. Within the divisions we decide which problems they shall work on. This is very clearly defined. The division has to pay for part of it. The rest is paid from the head office. The company has to push the divisions to think for tomorrow, because the divisions think primarily of their result for the year. They have different problems. Therefore, a part of this money is paid by the company. They are also pursuing some independent research where there is no division behind it. This research is mainly concerned with general future trends.

Aspray: Was it in the mid-1980s that Daimler Benz came in?

Treptow: Yes.

Aspray: Did you notice any major differences in the advanced research? Was it operated or managed differently after they bought AEG?

Treptow: Not really. Before they acquired AEG, Daimler Benz was more or less a motor car and truck company. AEG was an electrical engineering company. So at the beginning it was a new world for both sides. But then the decision was made to bring all these research laboratories together under one roof. A lot of synergies were found in computer programs that they had, which we could never develop because of the expense.

I can tell you about one problem we had with our new circuit breakers. It is a problem in the arc with this hot gas. This hot gas goes through channels inside the breaker. We asked the people from the motor development organization if they have ideas how to calculate those things. They said, "Oh, we can do that." They had a big computer program. It took them a fortnight and then we had the results. Otherwise it would have taken us two or three years.

Aspray: Suppose those people on advanced research come up with what looks like a promising idea. What is the next step towards product development? How is that achieved in the company?

Treptow: If it is a brand new thing that nobody has thought about, I don't know. For example, there are many ideas about batteries for electric automobiles. In the beginning the company got an external partner and we did the research work together. When the research showed that this was a very effective battery, much better than a normal car battery, there was the question, where to manufacture it? Inside AEG there was a division which was not so far away from the technology involved with developing batteries. Then this division got the order, together with the research people, to start a pilot factory.

Aspray: Do you have a specific division that is set aside for developing manufacturing processes for new products?

Treptow: Yes, we have.

Aspray: Let's say your field people come to the research and development division and say, "We have a problem with this, would you work on it?" Your research and development people come up with a solution involving a slightly modified circuit breaker. Before that goes into production, some additional work needs to be done. How does that get done?

Treptow: The research people never come with the finished product. They stop much earlier.

Aspray: You mentioned earlier the synergies that took place in the research division under the current situation. When Daimler Benz purchased AEG, what were the reasons? What were the anticipated synergies in bringing them together?

Treptow: I don't know. But at the time the newspapers suggested that businesses limited to trucks and cars may run into problems in the next twenty or thirty years. A revival of the train business was under serious consideration. We have a big boom in trains at the moment in Europe—trams, metro systems, and people movers. These technologies are technologies of AEG. Therefore if you look at the Daimler Benz group now, the Daimler Benz group has all systems dealing with transport. They do cars, trucks. They do trams, metros. They do high-speed trains. They do airplanes and space. They even do traffic systems. We think there will be some revolutions in this field to combine all these cars and trucks and planes and bring them into a better, more efficient system.

Aspray: How did organization or management philosophy change with the coming of Daimler Benz? How did the corporate culture change?

Treptow: Now there is an interesting discussion. Some very old companies came together. AEG is 110 years old. Daimler Benz is 105 years old. Then Dornier came to the group—a very special company with a real family touch. Then came the BM group along with the other airplane companies. There are quite different cultures in those groups. It was a big discussion, what is the best? Should we change it all to one new culture or keep the cultures in parts and try to weave them together into a multicultural company? The result is that car people are still car people. We are still electrical people. And there are others. But there is a feeling that we are all in one boat. I think that it is a big success. The last eight years, we have really grown to know each other. We are now on the level where we can phone each other if we know where the people are. In my opinion, it works much better than I thought it would.

Aspray: Do people move between the companies?

Treptow: Yes. Our boss, Mr. Stoeckl, comes from Mercedes Benz. And a colleague of mine went to debis as a marketing man. So, yes, there are some changes and some job exportation.

Aspray: It seems a little harder for technically oriented people to move across than it does for financial people.

Treptow: Absolutely. It's nearly impossible for research people or for development people. It is particularly unlikely inside AEG because of the degree of specialization. If you have a specialist for PLCs and bring him to transformers, he has no chance. That is a real problem but you can change some of them after a few years. That is what we are doing. If somebody is twenty years in one business, you can't change him. But with a manufacturing man, it's much easier. At the end, it's not so important if you manufacture cars or refrigerators or motors. There are many things that are very similar. But research is different. For finance people, general managers, or controllers it is easier.

Aspray: Besides research, what else has been centralized?

Treptow: The way of financing. Our bank is Daimler Benz.

Aspray: Is the supply system?

Treptow: Yes and no. Sometimes we find some steel or sheet materials that more than one division can use. But otherwise it is not centralized.

Aspray: Do you centralize accounting or marketing?

Treptow: Not in the whole group. The marketing people may attend some meetings to talk about what is going on in other divisions. That happens, but it's not centralized.

Aspray: Does each of the companies within the organization act as a cost center itself?

Treptow: They act within business units. They can be fairly small if they are highly specialized. We have a very successful company specializing in lighting systems. This is a company with 200 to 250 million Deutsche Mark turnover. It's a real company with all that you need including development, manufacturing, sales, everything.

Aspray: Are the companies given very strong independence in their action?

Treptow: Yes. I think so.

Aspray: What kinds of decisions are made centrally?

Treptow: We have to remain separate. As you know, the Daimler Benz group has four groups. There is AEG. There is Mercedes Benz, which is cars and trucks. There is Deutsche Aerospace, which is aerospace and defense. And there is debis, which is financial services and insurance. Each of these four sectors has its own board. That is one level. The chairmen of these four are also members of the Daimler Benz board, where Mr. Reuter is the chairman. There are situations where we have to go to the main Daimler Benz board. Otherwise, we make a lot of decisions inside our own board.

Aspray: In your particular area of business, do you and your seven competitors enjoy a fairly stable market? Have they been in this business for a long time? How do the market and the players change over time?

Treptow: This is a very unique aspect of my business. There are big barriers to entry for newcomers. There have been no new players the last ten years. There have been a few joint ventures but no real newcomers from outside.

Aspray: Twenty years ago, were there many more players in this game?

Treptow: Yes.

Aspray: Did most of the shrinking of the business come from corporate mergers or from people leaving the business?

Treptow: Both.

Aspray: Did AEG buy up a number of competitors?

Treptow: Yes.

Aspray: Over time AEG has bought a few companies and merged them into it. How do you make that business decision? What factors do you look at?

Treptow: For many years AEG was a German company with some exports to other countries. But we had some parts on the map where we had no chance to get in. So we acquired companies in those parts of the world. We acquired a company in Italy. We acquired a company in Belgium. We have joint ventures in India and Mexico. These companies are tight, fairly strong, and tied to the general business. A lot of technological thinking is done here. But we do the work locally. We do the business locally.

Aspray: So you are not purchasing companies because they have a product you would like to have in your product line.

Treptow: That also can happen. In Italy there were both. They had very interesting, very fine products. Their products, however, are on a lower level than ours. AEG has a fairly high quality level and are in the upper region of price scale. But in many parts of the world, that is not always the way. Our high quality products may be too expensive.

Aspray: Is there much cross licensing of patents in this business?

Treptow: Yes.

Aspray: Let me turn to your everyday work. What are the most challenging things that you face in your work? What kinds of issues are the most difficult?

Treptow: This interview. [laughter] This is my first day after finishing my normal work in the company. From today I am retired. In AEG it was normal to leave the company with sixty years. I prolonged it a little.

Aspray: Before your retirement, what were the most sensitive decisions? Which ones took your greatest talent to handle?

Treptow: Looking at the markets and the technology. For instance, we had to determine production costs in Manila. You can't compare the cost there with the labor cost in Germany. But that is fairly far away. So you have enormous transport costs. But we did electronics products there.

Aspray: Because they are inexpensive to ship.

Treptow: Right. Since the changes in Europe and the collapse of the Russian empire, we have highly skilled workers just beside us with one tenth of our costs. One tenth and less than that. So we have cheap labor and good skilled people just beside us. This raises tremendous problems in Western Germany. We are always asking the question, what should we do with our cost level, with our factories? In Germany it is much more expensive and much more difficult to close down a factory than in the United States.

Aspray: Because of government regulations?

Treptow: Not only government. It also costs such a tremendous amount of money to get rid of a worker. You have to pay him to get rid of him. This is one of the main problems.

Aspray: What has AEG's initial response to this been? Have they started to close factories?

Treptow: At the moment we have a depression. So it is twice as hard as normal. We just closed three small factories in Austria. We also had to reduce employment here in Germany. It's a big problem. In recent years, those problems are much more important than the real technical problems associated with products. We spent a lot of money in research and development. So the problems we have to fight at the moment are cost problems.

Aspray: I think I understand this point. Let me concentrate, nevertheless, on some of those technological issues. What kinds of technological issues do you face at this management level? And what kind of technical knowledge do you need to have to do your job?

Treptow: That is a difficult question. I am an electrical engineer. But my colleague, who is responsible for the automation side, is not an engineer. He is doing a good job. So it is difficult to make generalizations. But still, I think that a good technological base helps you a lot in your job. You can talk to all the people in the factory and in the laboratories. You know what you are talking about. You don't know every bit you are working with, but you know what they are doing. You can talk to them and you understand them. That is much easier than if you have a commercial or legal background.

On the other hand, in my career I often saw it would be advantageous to have a better education in finance, in commerce, or law. I see in our universities there is a gap. We are educated more or less as research people at universities. But I would say it is very important for an engineer also to have a financial and commercial background. You can say I got a lot of those things during my career. I know what a balance sheet looks like. But I have no chance to discuss every point with the specialists. I think there is a gap in the education of engineers in universities in Germany. Maybe in the last year it changed a little. We have a faculty where you can study engineering plus a commercial course.

I have always been glad to have this electrical, technological base. It helped me a lot. I did some research and side development. It was a good experience in order to work together with real workers. I never had problems talking to them.

Aspray: Do most of the people that reach the top levels of management come from one particular kind of background?

Treptow: No. If you look at the big industrial companies in Germany, for instance, the chairman of Siemens is a lawyer. At AEG, he is an economist. You find in recent years a change from engineers to financial people or lawyers. Maybe it's the same in the states. I'm not so sure. Jack Welch [CEO at General Electric] is not an engineer.

Aspray: But maybe it's at the divisional level and not at the senior level that there are more engineers in management positions.

Treptow: This is correct. On our board, we have seven members presently. Out of seven, two are engineers. But if you go a step down, you get more and more engineers. There we need the technical knowledge. So there you have engineers.

Aspray: These engineers in the middle level management positions also need to have some management skills. Does the company have a program for doing this?

Treptow: Yes. A real tough program.

Aspray: Can you describe how it works?

Treptow: Starting from the level of foreman they have to go to different seminars. We have a lot of such seminars. We always try to train those people, not in special techniques, but rather as a team. We train the engineers on the commercial side and also on the legal side with the workers. That is the lower level. We train the higher people in making presentations. But we are not as good as Americans in this field. If you are ever in a presentation from our American managers, they are great.

We have what we call the module system. The beginning of the module system is how to talk to people and how to manage a group. Then there are other modules for technicians or for commercial people. We do a lot in this field. The higher middle management also have the training together with Daimler Benz. But we always try to train them in teamwork. And we have an exchange between people in the headquarters where we do things for the whole company and people in the factories so that they can have an idea about both sides.

Aspray: What about going outside the company? Do you take advantage of things that the universities offer?

Treptow: Yes.

Aspray: Are these on the management side, the technical side, or both?

Treptow: Both. Here in Germany we have a system that is called "university seminar." That is a system which is sponsored by industry and universities. It is a six-week training program at the university.

Aspray: It seems that all Americans these days are interested in the Japanese. You mentioned before that in one part of your business you have strong Japanese competitors. Is there something special about them as competitors in your industry?

Treptow: Yes. We have a big competition from the Japanese in many fields. But here in Europe, they are not in all fields. They are not strong in power distribution in Europe. But they are very strong in automotive, as you know. Also in TV. But we are in big competition in certain markets. For instance, we are in intense competition in Korea. We compete over the high-speed train. We know a lot about Japanese thinking and how they manage factories. Our people have been there. We work together with them.

Aspray: From your point of view, how does their management differ from German management?

Treptow: Completely. The Japanese are quite different people. Their thinking is quite different. The whole culture is so different. In the past Europeans tried to copy them and it was not always successful. We copy some things. But not all of what the Japanese are doing is good for us. On the other hand, an engineer in Japan works thirty percent longer than a German engineer.

Aspray: Were there people above you in the company that were looking out for your career?

Treptow: Yes.

Aspray: Is that the way in this company?

Treptow: Yes. Also, if there is an opportunity and you are there, you have a chance to get the next job. If you are not there, you missed that chance. So it's a lot of luck in your career. In the end, I don't know if it is always getting to the top that really counts. That's a big question.

Part II
Japan

Chapter 5
Katsutaro Kataoka

About Alps

Alps, currently one of the largest electronics companies in the world, began humbly in 1948 when Katsutaro Kataoka borrowed fourteen hundred dollars from his family to establish Kataoka Electric Company. Kataoka originally produced only simple components such as light switches and variable capacitors. But beginning in the 1950s, the company began investing more heavily in research and development, opened new factories, and developed a wider variety of high-quality components in response to pressures from its increasingly technological and diverse clientele.

Since its founding, Alps has exercised an unusual and effective corporate strategy, calling for specialization and reliance on specific technological capabilities in electronics components manufacturing. Because Alps has traditionally been concerned only with supplying first-quality components to its clients (which include, among many others, Apple, IBM, Honda, General Motors, Goldstar, Matsushita, and Hitachi), it has been able to take advantage of greater economies of scale while avoiding direct competition with its clients.

Although Alps has limited its role to secondary manufacturer, it has expanded its operations strategically and consistently throughout its history. In 1964 the company acquired Tohoku Alps as a subsidiary and changed the entire company's name from Kataoka Electric Company to Alps. During the mid-1960s Japan's consumer electronics industry experienced unprecedented growth. Suddenly Alps components were being incorporated into thousands of products. Alps became particularly successful in radio and UHF television tuner technology. Secondary manufacturers like Alps made Japan's export-led boom in electronics possible by increasing production to meet their clients' skyrocketing needs.

By 1970 Alps was the largest independent component manufacturer in Japan. In the face of depressed exchange rates, Alps avoided making price concessions to primary manufacturers by developing special components and beefing up research on new end-products. In the same decade, Alps entered a joint venture with Motorola to manufacture car stereo equipment (later known as Alpine Electronics) and semiconductors. IBM and Apple joined the long roster of Alps clients in the 1980s when they contracted with Alps to design their computer keyboards. By 1985 Alps was the largest manufacturer of floppy disk drives.

Although Alps continues today to devote most of its resources to serving as a secondary supplier, President Masataka Kataoka is implementing plans to reduce Alps's reliance on secondary manufacturing. In 1991 Alps unveiled its Central Research Lab to unify the Alps research and development (R&D) network and enhance product development in computers, communications, and car electronics. In response to electronic globalization, Alps has developed a four-part corporate organization with bases in Japan, Asia, America, and Europe in order to achieve smooth and steady overseas development. With foreign subsidiaries in sixteen foreign companies already, Alps is looking next to expand its presence in the People's Republic of China.

Katsutaro Kataoka

Place: Tokyo, Japan

Date: February 17, 1993

[*Note:* Mr. Kataoka speaks in Japanese, and his words are interpreted by Hirotoshi Okamura, Manager of the Patent and Legal Department at Alps Electric. Also present at the interview was Professor Yuzo Takahashi of Tokyo University of Agriculture and Technology.]

Aspray: I want to find out what you attribute to be the factors that allowed your company to be so successful as a start-up after the war.

Kataoka: I graduated with a degree in mechanical engineering from Kobe University. Then I entered Toshiba Corporation in its communications department. That was in 1937. After nine months of experience with Toshiba, I went into the military service. I served in the military for five years and ended my service as a lieutenant in the wireless communications troop. In 1943 I rejoined Toshiba. On August 15, 1945—that was the last day of the Second World War—I left Toshiba.

Aspray: Why was that?

Kataoka: Toshiba was too big for me. After I left Toshiba, I assisted with a friend's company for a short while. Then I established this company on November 1, 1948.

During my time with Toshiba, I was in charge of certain machine work as well as making variable capacitors and switches, which were designed for use in the radio transmission systems of airplanes. Based on those technologies, which I learned at Toshiba, I started to make rotary switches and variable capacitors at my new company. Five years after I established this company, black-and-white TV manufacturing began in Japan.

Aspray: Was it difficult at the time to start a new company?

Kataoka: Yes. At first, I had financial troubles. Japan was in a severe economic depression at that time. Also, the electronics market was small then. Companies such as Toshiba, which now have very large operations, were still relatively inactive due to the depression. Such companies were engaged in the process of deciding how to develop and expand. My only customers at first were amateurs, who were making radios by themselves. So, it was indeed a very small market.

Aspray: Was it difficult to build these technologies, these switches and resistors?

Kataoka: From the technological standpoint, it was not so difficult to make those types of products, but the market was very small. As a result, I encountered much difficulty from a business point of view.

I experienced severe business obstacles for about three years, but then the Korean War started. That helped my business a great deal. Let me explain. At that time, we had radio broadcasting services in the private sector, but we did not have any such broadcasting which used shortwave. That was only used by the military services.

We had many Koreans who were living in Japan at that time, and they were quite eager to listen to radio broadcasts originating from Korea, Seoul or Pian Yang. They liked to use shortwave to receive information about their home country from Korean broadcasting systems. Those Koreans were just waiting to buy our components to make shortwave receivers. By serving the Korean people, we were empowered from a financial and technological standpoint. Building on this foundation, I decided to start manufacturing TV tuners. This product was quite new to us and, I think, also new to Japan.

At that time I said, "If we are only making so-called discrete-type components, we do not have any future." So I decided to manufacture TV tuners.

Aspray: That is a much more difficult technology than the ones you had been producing before. How did the company manage the increase in technology?

Kataoka: The introduction of black-and-white TV was quite new to the Japanese electronics industries. At that time, there was little or no basic technology for such a TV. Because of that lack of technology, we were struggling to determine appropriate specifications for the TV tuners.

We started to do business with Motorola as well as doing our business with Japanese electronics manufacturers. During the course of our business with Motorola, we learned and progressed a great deal from a technological standpoint. We already had previous business with Motorola for other products, but then Motorola also began to give us orders for tuners. At that time, Motorola gave us a great deal of assistance.

Previously, TV set manufacturers made their own tuners, and their tuner specifications were not available to us. However, Motorola provided us with such specifications. If I remember correctly, Motorola gave us test equipment to use that cost more than one hundred thousand dollars. We are still grateful to Motorola for this assistance.

In 1957 I was sent by Japan's Productivity Center Institute to visit the United States. They sent members to the United States for study, and I acted as secretary to those members. I stayed in the United States for about fifty days and visited fifteen factories. The United States government and American companies supported this endeavor and gave me the chance to understand the current status of American factories. I had access to those companies—to see their facilities and to engage in discussions about their business.

Aspray: Those circumstances were very different from a later period of time.

Kataoka: What do you mean?

Aspray: Later on companies were more protective of their factories.

Kataoka: It is my opinion that the United States changed its attitude regarding Japan and the electronics industry. The United States and Japan had a long-term relationship with the United States acting as teacher and Japan acting as student. The Vietnam War proved to be a turning point. After that, the United States concentrated almost exclusively on the military industry and forgot about the consumer electronics industry.

Aspray: I have a couple of questions about the Motorola relationship. Do I understand correctly that Alps was producing the tuners not only for the Japanese market, but also for Motorola's entire market? And secondly, what do you think were the reasons that Alps was chosen? There were manufacturers in many different places. Why select Alps?

Kataoka: The answer requires some explanation of our relationship with Motorola. In the very early part of the 1950s, RCA and General Electric had purchasing offices in Japan. But Motorola was looking for the same opportunity to start business in Japan. So Motorola decided to follow IBM, NCT, GE, and RCA by opening its own office here in Japan. They established a purchasing office in Tokyo in the early part of the 1960s. Our company sold switches and variable capacitors to this office, but those were basically used for radio sets. Because of the differences in the measurement systems of the United States and Japan, it took about two years to get final approval for Motorola to use our TV tuners.

Aspray: Within the company?

Kataoka: Yes. I have very memorable stories about my dealings with Motorola. At that time, the purchasing director was Mr. O'Brien. Unfortunately, he has since passed away, but we became very well acquainted back in those years. I had tried to get approval for our tuners from Motorola for two years, but finally I decided to give up. I contacted Mr. O'Brien to convey the message, "We simply give up." Mr. O'Brien told me that maybe we were running the equivalent of a three thousand-yard race, but we were approaching yard number 2999. He said that we had only one yard left to go. Why not try to complete that last yard? Mr. O'Brien then lent us test equipment of Motorola that cost about one hundred thousand dollars to help us overcome correlation problems and make quality tuners.

At that time there was no good measurement or test equipment for high-frequency products, for example, television tuners and so on. Another obstacle was the difference in the measurement systems. Your system was being measured in inches while we were using the metric system. This caused us further difficulty.

At that time, however, most Americans had no bad feelings toward the Japanese. Americans were very open-minded and helped me very much. That time was probably America's apex.

Aspray: I take it that the tuner business expanded and there became other customers in Japan as well as in the United States.

Kataoka: The tuner business was indeed expanding, but at that time most of the Japanese TV manufacturers were making TV tuners by themselves—through in-house production of the tuners. From our company standpoint, business had not expanded in that area.

Aspray: What were the major products at the time?

Kataoka: Switches, mainly rotary switches. And variable capacitors.

Aspray: So it was a continuation of the original business of the company.

Kataoka: We had the number one share for those product areas. Those two products, rotary switches and variable capacitors, helped our gross sales and profits. In other words, those two products were our core products.

Aspray: By the 1950s, there must have been a number of other competitors in Japan in these businesses. What allowed Alps to have the number one market share in these areas?

Kataoka: As far as the rotary switches are concerned, we did not have any real competitors. As for the variable capacitors, we had at least twenty competitors. Most of them were making variable capacitors for radio receivers. The introduction of the superheterodyne circuit for the radio receiver probably occurred in the early 1950s. At that time, we were making variable capacitors of the same high quality that was necessary for airplane radio transmission equipment. So, our high-quality variable capacitors were very much suited to new products such as the superheterodyne circuit.

Aspray: Where did the business go then?

Kataoka: After that came UHF tuners. We started our UHF tuner business in 1962 and two years later we concluded a technical assistance agreement with General Instruments Corporation (GIC). We were able to introduce technology from GIC. Around 1962, through legislation passed in both houses of the United States Congress and FCC rulings, all tuners needed to carry all channels. From that point on, in other words, all tuners exported from Japan to the United States had to have two functions: VHF and UHF.

Aspray: Were the Japanese manufacturers ready to provide that product right away?

Kataoka: No. Around that time in Japan, no TV broadcasting services were using UHF channels. Therefore, no one was prepared to manufacture UHF tuners. Fortunately, UHF technology was based on the variable capacitor. Initially, only Alps was in a position to manufacture such UHF tuners.

Aspray: In order to do this, did Alps need to build new factories, hire new kinds of specialized engineers, or start a research and development operation?

Kataoka: Yes. Japan did not have any technology regarding TVs or tuners, so we had to learn these areas by ourselves. In the 1950s most of the Japanese TV manufacturers had some quality factory engineers and mechanical engineers for TVs.

However, there was also very big support, as I mentioned before, provided by Motorola and General Instruments. Before that, the most helpful company to the TV business in Japan had been RCA. They helped Japanese TV industries very much.

During the 1960s the head of RCA's Tokyo office helped the electronics industry in Japan. He gave us three principles which have been applied to Japanese component manufacturing ever since. First, quality. Second, delivery. Third, cost. Still we continue to adhere to these principles throughout Japanese component manufacturing. RCA also granted us a patent license for TV technology and provided us with technical assistance.

Aspray: How did the business expand from that point?

Kataoka: We started a business here in Tokyo at this point. This building we are now in used to be a manufacturing facility. In 1960, we expanded and built our Yokohama plant. At the Yokohama plant we had, at the most, 2,500 people. Starting in 1964, we started to build more factories in northern Japan.

Aspray: Why was that?

Kataoka: At that time, areas in northern Japan were not yet industrialized. They were basically agricultural in nature. However, there were many high-quality laborers there. Young people in these areas did not desire to enter the agricultural business and were moving from northern Japan to Tokyo.

Until 1964 or 1965, industries tended to locate in urban cities: Tokyo, Yokohama, Nagoya, Osaka, and Kobe. But, around that same time, businesses started to shift the location of their production facilities from the cities to local, rural areas. As for us, we were growing at an annual rate of more than 20 percent. By the 1970s, we were able to completely shift our production facilities to northern Japan.

Aspray: I assume that land was less expensive there and it was also more available so you could expand if you needed to.

Kataoka: Your assumption is correct. Another main issue was capital. In the 1960s, most of the component manufacturers started to list their shares on the stock exchange market to raise funds from the open market.

Aspray: So the company had been private before this and then it became a publicly-held company?

Kataoka: Yes, I think that it is quite an important point to understand Japanese component manufacturers from the viewpoint of capital. According to my past experience, most of the component manufacturers all over the world at that time were privately held companies and family-type operations.

Most American components manufacturers were advanced very much ahead of any other country's component manufacturers. When I visited the United States, I saw some big component manufacturers which entered the stock exchange market. I learned the American component manufacturers' way of operating business in terms of how to raise funds from the market. Because I had a shortage of money, I decided I would also open our company shares to the stock exchange market. Today 80 percent of electronic components are manufactured by component manufacturers who are listed on the stock market.

When we started the business, we had insufficient technologies, so we had to introduce technologies from the United States. A second issue was capital and entering the stock market. When we became a publicly held company, we could get money from the market. Using those funds and the American technology, we were finally able to expand our business.

In the early part of the 1970s, most Japanese component manufacturers were recognized worldwide as very large manufacturers. American technology was still very powerful, and most Americans could not see that in a short time Japanese component manufacturers and electronics manufacturers would become huge, affluent companies.

Just ten years ago, I began questioning my American friends, all executives in American corporations. "You taught us so many things. For example, you told us the first priority should be quality. But you didn't apply your principles to your operations." So I ask, "Why?"

Aspray: What answer did they give you?

Kataoka: They agreed with me. In my private opinion, what causes the disparity between their indicated agreement and their actual behavior is that American executives and their corporations are driven by short-term output goals. In contrast, most Japanese companies focus on long-term solutions and goals. Another factor is that Japanese companies have followed a system of lifetime employment and have traditionally maintained the same management personnel for many years; in this way, Japanese companies try to be, in a manner of speaking, eternal. In contrast, American companies often have personnel changes at the executive level. Sometimes this causes difficulty in negotiations.

Aspray: I understand that the company changed its name in 1964. What was the reason for that?

Kataoka: At that time, I had relations with Motorola and relations with the General Instruments sales force. This is very funny, but some people who speak English cannot properly pronounce my name, Kataoka; they pronounce it "Carioka." Someone pointed out that we had a very good trademark, Alps. So, why not just use Alps as the company name instead of Kataoka? I agreed. I was lucky because we were successful in registering the name Alps all over the world.

Aspray: It just happened to be free.

Kataoka: That's right.

Aspray: How did you manage such tremendous growth in the 1960s? What were the management challenges to sustaining that growth?

Kataoka: I had experience with Toshiba working for a huge organization. I also had experience with the military's large structures and organizations. I was lucky to bring those two experiences to this company.

I have also been looking at three companies over the long term: Motorola, General Instruments, and Nortronics. Only Nortronics is still making magnetic heads—even though Nortronics was purchased by an American company. I could learn from those three companies and their very powerful American engineering activities.

Twenty years ago, I established one building called the Alps Training Center with the understanding that it was mainly for Alps employees who were university graduates. They could learn many things there even if such knowledge may not be directly or immediately applicable to our business. I decided to give them the opportunity to acquire more skills while working for our company.

Aspray: What kinds of topics were taught in this training center?

Kataoka: Mainly we provided training on management skills. We were also training instructors in this area. We then sent those persons to instruct the divisions throughout our company's operations. These instructors could then discuss our policies with persons who are working in our many divisions.

In the past ten years, technology has advanced very rapidly. In order to sustain the rapid transfer of technologies, we also decided to send our engineers to laboratories or institutes controlled by professors to exchange information and to have coordinated activities at the university level.

Aspray: I don't know much about the Japanese system. In order to go to the universities did your company have to supply equipment or funds to the university to get knowledge?

Kataoka: We did not spend a significant amount of money, and we did not give universities any kind of equipment. However, if a group within the company engaged in activities under certain professors, then those groups needed to support such professors from a financial standpoint—perhaps through small donations.

The government promotes decentralization of industrial activities outside of Tokyo. So the government provides support to establish certain technical coalitions. For example, we have been working together on R&D activities with Tohoku University to develop technologies.

Sometimes Americans may misunderstand these situations, but they must understand that Japanese private companies are not directly supported by the government. The technology policies in the two countries are quite different. We are just following the government policies for decentralization. We work together with the universities and local governments.

Aspray: Coming back again to ask a little more about this rapid growth, it seems that you faced a whole series of challenges. You want to have enough factories and personnel to be able to expand, but you don't want too many so you can keep your overhead low. You have people rising rapidly in the company but maybe they go too high, beyond their talents. You have problems with your suppliers, and so on. Can you give any advice about other lessons that can be learned from your experience here at Alps?

Kataoka: We are always having small troubles or big troubles because we are managing a large technological company. I can explain some big troubles which occurred during my management. In 1974, when we had a depression in Japan, we asked 3,000 of our 11,500 people to leave the company. Roughly 30 percent of the people were asked to leave. As you know there is quite a big difference in the employment systems of America and Japan. We have been practicing a sort of lifetime employment system. Labor here is also somewhat protected under the law. In the United States, it is much easier to lay off people. But in Japan we could not reduce our work force so easily.

A second big problem was the appreciation of the yen. For example, many fluctuations occurred in 1979.

A third major problem is one we are currently having. We are currently facing recession. It is the same all over the world. In Japan, we call it the collapse of the bubble economy.

Aspray: Did going off of the gold standard cause a problem for you?

Kataoka: Yes. When we started business, there was a fixed exchange rate of 360 yen per U.S. dollar. With the fluctuation, our business is made more difficult. Now the exchange rate is approximately 120 yen per U.S. dollar.

Aspray: At some point did Alps become the supplier of components to the large Japanese manufacturers of systems? When did that happen?

Kataoka: In the middle of the 1950s—either 1954 or 1955. The EIAJ (Electronic Industries Association of Japan) has been keeping certain industrial records on growth by companies. Alps's annual growth was consistent with this industry data until recently. Recently, Alps's overseas production has increased, and such production is not reflected in the EIAJ data. As a result, the EIAJ data must be adjusted to reflect reality.

Aspray: It seems to me that Alps has been extremely successful in secondary manufacturing, component manufacturing. There are fundamental tensions there because the systems manufacturers who buy your components have a great deal of power over the companies. How was it that Alps was successful? What strategies did they use to make sure that they could continue their growth and profitability while they had to deal with these powerful companies that were much larger than they?

Kataoka: We have been upholding the principles suggested by the head of the RCA Tokyo office, as I told you. Number one should be quality. Second should be delivery. Third should be cost. We have been relying on these principles very much. We do not primarily manufacture products according to the drawings given by manufacturers. Instead, we mainly generate such drawings and make products according to our own designs.

Our engineers work in close cooperation with the manufacturer's engineers, so we have close communications with their engineers. It is true that our customers are very powerful. Intermediate between systems and component manufacturers, we have a formal organization called EIAJ (Electronic Industries Association of Japan), as I mentioned before.

Each company also engages in very useful, regular, informal meetings with key members of its suppliers. We call these cooperative group meetings.

Set manufacturers also belong to certain organizations. In those organizations, most of the component manufacturers are invited as members, so that we can exchange our opinions.

Aspray: Does the industry organization have any real power or are they just a persuasive body?

Kataoka: The ultimate power resides in each individual company. By using the organization EIAJ, however, we can submit our companies' opinions. Under the aegis of EIAJ, we have at least three hundred committees in which we discuss technical issues and other relevant issues. I formerly was the vice president of the EIAJ. I was also the chairman of the component committee within the EIAJ. Through these organizations, we have good communications with our customers, that is, the big manufacturers, and they will listen to our opinions. They may not always change based on our opinions, but at least they listen.

Your government and also the leaders of American corporations have said that Japanese companies have *Keiretsu*. Do you understand *Keiretsu?* In my opinion, this is not true. We are living in a free market, and we are in a very competitive situation.

We have been dealing in this industry by using our own processes and technologies. But, we do have some group meetings to assist us in communicating with our customers. These are associations of suppliers sponsored by the suppliers' customer. These associations basically consist of assemblers, discrete component manufacturers, and raw material suppliers. In fact, we have several competitors within these same group meetings, and we have very free exchanges of information both horizontally and vertically.

At this point, I would like to mention that I believe there is one big difference between various industries. That is the life cycle of the industry's products. In the automotive areas, the cycle is five years. But in our industry, electronics, it is only one year. Because of the long life cycle for products in the automotive industry, this can lead to some very close R&D activities between auto manufacturers and their suppliers. When Americans look at these activities between auto manufacturers and suppliers, they regard those activities as inappropriate. But in my opinion, that is a misunderstanding of the activities.

We do have business with auto manufacturers, a dashboard business. We have been supplying BMW in Germany with certain air conditioning control panels, and now our product will be used for the 1995 model year. We are still in the development stage at this time.

As you know, Alpine Electronics Inc., a subsidiary of Alps, has business with Honda. At least twice a year, Alpine and Honda have big technical meetings where the discussion is on products to be sold two or three years into the future. We are not discussing current products. Further, despite these long-term discussions, Honda is not obligated to use our company for future activities. If those discussions fail, our activities in two to three years will all be destroyed, because Honda might choose to use another supplier. In my opinion, most American companies are discussing Japanese companies' activities by looking only at the very nominal or surface appearance of these activities.

Aspray: Business theorists say that one of the strategies used by components manufacturers is horizontal diversification. That is, you take the component you are manufacturing and start broadening your line so that you don't have to rely so much on one business relationship. Instead, you have many relationships with many different industries. Was that a strategy in Alps, and did it succeed?

Kataoka: There are two methods of engaging in the components business. The first involves acting as a subcontractor, which is making products from drawings that come directly from manufacturers.

Alps is a good example of the second. We are primarily doing business based on our own drawings and designs. Because of this, we are capable of diversifying our business. We are expanding the business in various ways. For example, we have added components within the automotive product field.

But we are not increasing the number of variations on particular components. Instead, we are trying to standardize our components so that they can be used in a diverse range of final products. The reason we can do this is that we are doing business according to our own designs.

We used to have more than ten thousand product lines, but recently most of the Japanese component manufacturers have tried to reduce the number of component varieties offered. For example, we have reduced from more than ten thousand to five thousand product lines to reduce our overhead costs and manufacturing costs.

When I checked the code numbers for our products, the code numbers were based on the customers and their products. But there were still over three hundred thousand. That is too many. So we are now trying to reduce those numbers even more. We would prefer to reduce to half of these current figures.

The reason we have had this large number of products is that we were following improvements in integrated circuits. The newer integrated circuits allowed for down-sizing, and we ended up with so many tiny products.

Aspray: Can you tell me about some of your new businesses of the 1980s? I know, for example, that you have become a major manufacturer in floppy disk drives and computer printers. Could you tell me how you got into that business and how it grew?

Kataoka: We started the floppy business with Apple Computer first. We established our own technologies to make thermal transfer printers. Then we built our market by developing those technologies. The application has been mostly for word processors.

Actually, we made dot matrix printers first but we gave up because we were not successful in that business.

Aspray: In most cases were the floppy drives sold to a systems manufacturer? Were they then put into somebody else's machine?

Kataoka: Yes.

Aspray: Has the company tried to get into primary manufacturing, to produce finished products? Has that been a goal of the company?

Kataoka: We probably have the technology to make a complete set of products, but in my opinion, we do not have any sales power to send those products to the market. We established Alpine to send products directly to the customers in a specialized market. Alpine's main business is in specialized products for use in automobiles—especially with Honda and other auto manufacturers. Alpine is successful in the automotive business. But it is quite a different situation when you get into general components manufacturing.

Aspray: Yes. It just seems so totally different from the kind of business you have been in. Are there any parts of your business that we have not talked about that you would like to mention to me?

Kataoka: No, but I do have several additional comments. First, I would like American management to return to their old style of corporate management. They used to give so many things to Japanese industries. So we want to know, "Why don't they remember the past, and come back?" There have been no U.S. manufacturers of radio sets, color TVs, VCRs, tape recorders, and video cameras for some time. Even if engineers graduating from the university want to get into this market or these technological fields, there are actually no openings for these engineers in the United States. They have nowhere to go.

Second, American industry should keep a very close eye on Asian countries and in particular on mainland China's activities, especially the labor force and the economy.

I have these two points for the Americans, and I have constantly been telling my American friends these two points. In my opinion, it is not best to attempt to solve problems through discussions with government bodies.

I should also mention labor issues in the Japanese electronics industry. Of course, we have some experience with labor issues, such as strikes and other action the labor union has taken against the company. But the electronics industry as a whole thus far has had a very good relationship with labor. In my opinion, that will not be changed in the future. In the future, I believe labor will not be an issue for the electronics industry, because the thinking by most top management personnel at these electronic companies parallels the thinking of those leading the labor unions. Each side understands that it must grant concessions. So, in my opinion, the heads of the labor unions will continue to have a small role in the future.

I believe that the Japanese electronics business has learned many things from the United States electronics industry. When I visited a person from General Motors who has studied manufacturing methods, I mentioned that we have learned so many things from the United States and that Americans should do the same from our operations. To promote such study, we opened our factory to a delegation from General Motors.

Okamura: Mr. Kataoka has shared this opportunity with a total of eighty young engineers and middle managers from General Motors. I think he has also been contributing, not only to the company, but also to industry and to Japan itself. In fact, he was honored by the government and received the Second Class Order of the Sacred Treasuries in 1983 because of his contributions to Japan's industries.

Aspray: Very nice.

Okamura: In my opinion as an associate of Mr. Kataoka, he has contributed so much to the company and also to the electronics industry. He learned many important things during his time in the United States under the Japanese Productivity Institute. He saw that so many American component manufacturers were corporations which stood independent from the systems manufacturers. He took this concept and reasoned that Japanese components manufacturers could also survive in the market as independent manufacturers.

He misses his friends in the United States. Most of his closest friends are retired from their companies or have moved away. Mr. Kataoka said that throughout this interview he has been experiencing some nostalgia for the past. Only one person is still active in the industry, Mr. Galvin from Motorola.

Aspray: Robert Galvin, yes.

Kataoka: Now our company is operated by the second generation of management, and I am now half-retired. [chuckling] There are very few persons who can speak about the background and origins of this part of the electronics industry. I am one of the few remaining persons.

Aspray: Yes, indeed. Thank you.

Chapter 6

Koji Kobayashi

About NEC

NEC Corporation was founded in 1899 in Tokyo as Nippon Electric Company, Ltd. Western Electric Company, the manufacturing arm of AT&T, designed the Japan-United States joint venture as a plant for its Far East market. They hoped to use NEC to establish a strong market presence in Japan, as they had in Europe. NEC initially was little more than a distributor of imported telephone equipment from Western Electric. However, NEC's size and focus changed drastically when it began receiving substantial orders from the NTT [the national telephone company] for the Japanese government-sponsored telephone-network-expansion program. This turn-of-the-century government affiliation was the first step in NEC's climb to its current position as the largest computer and telecommunications company in the world.

Under Western Electric's direction, NEC merged with the cable-manufacturing division of Sumitomo Densen Seizosho in 1923, enabling it to assume an important role in the production of electrical cables. NEC entered the field of radio communications in 1924 by providing radio equipment for the newly established government-sponsored radio network. A year later NEC became involved in electron devices as it developed electron tubes for its own communications systems.

By 1943 Sumitomo had assumed full control of NEC and renamed it Sumitomo Communications Industries. In 1945 the Allied occupation authority ordered the dissolution of large Japanese enterprises such as Sumitomo. Only public works projects kept NEC solvent during several difficult recovery years in the late 1940s. But in 1950 NEC began research and development in the field of transistors. Three years later NEC entered the consumer electronics market with its New Nippon Electric Company. By 1954 NEC had entered the computer field, and by 1963 it had embarked upon projects in the field of satellite communications.

After assuming the presidency of NEC in the early 1960s, Koji Kobayashi sought to maintain the company's strong growth rate by diversifying its product line more aggressively into new electronic product lines. He split the company's five divisions into fourteen, opened factories in countries all over the world, and promoted the integration of computer technology and communications expertise (C&C) in systems planning. The company enjoyed extensive international growth during the late 1960s and early 1970s under Kobayashi's leadership, opening factories in Mexico, Brazil, Australia, Korea, Iran, and Malaysia. NEC's foreign markets were largely responsible for the company's tripled sales figures of the 1980s. Today, the company's international operations in communications systems and equipment have expanded to over 140 countries. NEC is currently the largest manufacturer of semiconductors in the world and a competitive producer of broadcast equipment, satellites, and computers ranging from supercomputers to personal computers.

Koji Kobayashi

Place: Tokyo, Japan

Date: February 15, 1993

Kobayashi: There is a long history. You know the origin of NEC, Nippon Electric Company?

Aspray: I've read about it, yes.

Kobayashi: Sixty-four years ago I finished my study in the Tokyo University. This was a long time ago. But at that time there was no concept for the computer. Only of communications. Communications had many different parts in the Japanese industry. When I entered NEC in 1929, immediately after my graduation from the university, it was a subsidiary of Western Electric Company of the United States. I was thinking of a brighter future for communications technology when I was at Tokyo Imperial University. I decided to enter NEC because I understood that this was a joint venture with Western Electric and Japanese capital, 54 percent by Western Electric and 46 percent by the Japanese. I decided that NEC would be the best place for me to be involved with business, given my background in communications.

Anyhow, since then, the world has changed, many times. At that time when I entered NEC, there was not much concept for new technology. Everything, all that mattered, we were to learn from the United States. I knew how great Western Electric was, and I was impressed by its famous name. But we were not very happy to be overwhelmed by the name "Western Electric." To be honest, we could not do anything about that other than study, to follow the Western Electric way of technology.

In my early days here at NEC I was very unhappy because the majority of my work was just translating the drawings from Western Electric. But I wanted to do the research and development. That's why I joined. [Laughter] Then the Japanese social involvement shifted toward nationalism, and the government asked the company to develop Japanese military radio and acoustic systems.

Japan was very aggressive in capturing other countries, such as Manchuria. But we had no technology; no new technology had been developed. We just learned from Western Electric. Western Electric was a big company in the United States, a great company. Western Electric had the majority share of NEC. Even the management was directed by Western Electric. NEC was almost like an American company. So I decided: We are Japanese. We wish to stand on our own feet, even though it may be very difficult. We should do that. Otherwise, Nippon Electric cannot survive in the future. That was also my decision in determining the direction of my career in NEC.

Aspray: When did your vision of C&C, computers and communication converging, come to you? When in your career?

Kobayashi: I had no concept of the computer before the war or in the wartime. No engineer in Japan had such a concept. But we did have a concept of communications. We tried so much to develop a new technology for communications. Until the end of the war, we had little technology of our own. Japan had no original technology except for a few exceptions. As you know, we were completely defeated by the war and had nothing.

We had made some mistakes. We tried to follow only the technology that was developed by the Western countries. But I tended to think that if we confine ourselves to communications only, it's imperative to compete with Western Electric and Siemens from Europe. They had already established their own record of achievements in the area of communications around the world. Then what is left for a Japanese company like NEC? At that time, IBM was not so great as it is today. IBM was just coming up. So I was thinking: why do we stick to communications only? We need to explore new markets, new technology. The computer was a very new technology at that time. Very few companies were trying to establish their market in the computer. Not much difference between Japan and the Western countries. The situation was quite different from communications.

I thought we should enter into the computer business, but I did not know what the results of this would be. But, anyhow, the computer would give us a big opportunity in the marketplace. So, my thinking was, it's better for us to go into the computer business. Actually, we entered into computer research in 1954. Around that time computer technology was in its infancy, and NEC was early to market a best-selling product in Japan. In 1958 NEC completed NEAC-2201, an all-transistorized small-range computer.

Before the war, I visited AT&T, ITT, Siemens, and STC of England. All these companies seemed to have no idea about the computer. After the war, again, I had chances to visit these companies. They still had no idea about the computer. Against these telecommunications superpowers how could we compete?

On another aspect, although we made a very good start in the computer business, IBM was too great for us to compete successfully with. Other Japanese makers were also coming into the computer business. As the number one telecommunications maker in Japan and also a major computer maker in Japan, NEC had to compete with big companies of the world in both technology areas.

Entering the 1970s NEC firmly designed the direction it was to follow. It's information. Telecommunication is for transmitting information, and the computer is for processing information. NEC should specialize in information. Some overseas companies specialize in telecommunications, others in computers. No one covers both. Then, if NEC could cover both of these information-related technologies, it would be sure to obtain a comparative advantage over all these competitors. Why don't we have communications-computer? That was my ambition. [chuckling]

This idea of mine was first published in a very rough form in my article in an engineering magazine in 1959 in Japan. But it was a long time before this idea was announced to the world.

Aspray: I see.

Kobayashi: That is the concept of C&C. Computers and communications would ultimately come like this [Dr. Kobayashi brings his hands together] and integrate with each other. We developed quite a breakthrough, I think, a new market. This was the concept of C&C.

This was really the reason why NEC got into the computer business. A venture in this area, computer and communications, became very appealing. The real market, however, depended on the advancement of microelectronics. I brought the microelectronics business into NEC in the early part of 1960, and this finally led to the announcement of C&C in October 1977.

Aspray: In *The Rise of NEC,* you wrote that one of the differences between Japanese industry and American industry is that the American computer manufacturers came out of the business machine background, while it was the telecommunications background in Japan. How did you see the difference in the computer products in the two countries, based upon those differences in background?

Kobayashi: In the background, I studied much about IBM because IBM was the strongest company in the computer industry for a long time. At first I considered NEC's extensive relations with the foreign company ITT. ITT was the major stockholder—almost as though it owned NEC—for some time even after the war. (ITT kept its shareholding of more than 30 percent until as late as 1956.)

So after the war, when I started thinking about how to build up NEC, I felt it necessary to talk with the president of ITT, Harold Geneen. He was great. I met with him here in Japan and in New York. He had no engineering background, but both the computer and communications industries require complete understanding of the technology. Geneen's only interest seemed to be to make money in such businesses. I wanted to confirm what Geneen was thinking about these business areas, so I decided to see Mr. Geneen in New York to discuss this matter. At that time I already had made a commitment to a computing business. The C&C concept was already in my mind at that time. But I wished to check about the computer with Mr. Geneen. He did not talk much about such technology. He was thinking more about the financial side of the ITT operations. I felt I had to sympathize with ITT because I believed the business strategy of Mr. Geneen might keep ITT behind the times. It did.

As a company that was willing to spend big money for its computer operations, with the communications business as a background, at NEC we tried to enter into the computer business.

In the communications business, we made money. We earned money. We used these funds to support our computer business. I thought that without investing money for such a new area, we could not be competitive in the new market, that is, the computer business. I wondered whether this policy was correct or not. There was no other managerial decision than to announce that we were entering into the computer business in addition to our traditional communications business. Although NEC entered the computer business spending big money, that did not mean that NEC was slipping out of the communications business. We continued to keep very energetic activity in communications. But we also invested big money for computers. So I need to say that this was not really a matter of choice of communications *or* computers. It was C&C. So I was always talking about C&C. We were not only for this C, we were not only for the other C. We were for C&C.

I believe we could enjoy a synergy effect between these two mainstay business areas of telecommunications and computers. I wonder if computer makers with an origin in business machines could do the same thing.

I have another comment. Generally speaking, telecommunications engineers take a serious view of reliability and stability of systems and equipment. A grand communications network does not function unless systems reliability is maintained. Perhaps you know NEC was the first company in Japan to introduce Western Electric's quality control practices into a manufacturing shop. Both NEC and Western Electric are communications manufacturing companies. At these companies reliability is common sense where good component parts, good manufacturing processes, and good control are requisite. This sense of reliability was brought into other products we make—computers.

Aspray: Were you ever concerned that AT&T would take on such an approach? After all, they had had a strong history in telecommunications, of course, and they had done some work in computing before that.

Kobayashi: I did not recognize that AT&T or its subsidiaries had such an idea. I tried to make contact with AT&T executives and with Western Electric executives, but they did not have any idea of going into computers.

Aspray: I see.

Kobayashi: NEC and perhaps other Japanese computer companies always thought of computer-communications. At any time, from any place, anyone can talk to each other; that was our objective for advancing communications technology. With the entry into computer technology, our objective expanded: at any time from any place, anyone can access any kind of information, process it for their own purposes, and effectively utilize it.

We had no idea about the computer before the war, and during the wartime we were too busy with communications technology. After the war Japan was put into a very difficult position by the defeat in the war. But as time went on, I began to feel something about the idea of C&C as we call it today. My feeling is that IBM was great but they were only for computers. I would call AT&T great, but they were only for communications. I thought, why doesn't NEC go into such a business as C&C? No other company was thinking of C&C. So that must be a pretty dangerous idea, to commit the money and principal of the company to C&C. There are no ideas of offering shares. The idea was to commit ourselves to such behavior. Otherwise, we would never come into any power in the marketplace. That was the incentive for NEC to build up their business after the war. It was rather easy because we were not confined to communications. We were already in communications, and we could enter computing. Therefore, we could enter into C&C.

I left the working out of the idea to the engineers and scientists. The main objective at NEC had been continued growth—not commitment to a specific technology as such. There was a feeling that the company could enter into some new markets, which are built around C&C. At the INTELCOM Conference in Atlanta in 1977, I spoke extensively about C&C, expecting that some big corporations might decide to compete with us, or challenge our ideas, but they had no such idea.

After the INTELCOM Conference I also talked about C&C in Geneva and other places. I talked freely about the subject. The Geneva conference was attended by many technology experts and management executives from all over the world. I dared again to speak about C&C there. But nobody questioned me about my idea. I was convinced that NEC was put into a position where there was no competitor in the area of C&C. Anyhow, the announcement of this new concept was very successful. That's my feeling. [chuckling]

When NEC was limited to only the field of communications, it was difficult to make the company grow very rapidly. The computer is the newcomer in the field of C&C. In developing this C&C concept of the computer within the company, we developed many fresh ideas. If we had stuck too much with the communications, there would have been so many companies in the world competing with us. However, a learning process was required, all those under my employment and also my engineers asked me: "What is C&C?" I replied, "Oh, it is a new subject. I wish to learn about it from you. Please give your idea to me." Why did I insist upon C&C? It was because no other company seemed to enter the C&C business. Major Japanese electronics companies had no announcement for C&C. So C&C was monopolized by NEC. Fujitsu would not like to say "C&C." They used in newspaper [ads the phrase]: global computer and communications company. But mine is very simple: C&C. We were the only C&C company, with well-balanced stakes in computers and communications.

Aspray: As I read your book, it seemed to me that of all the Japanese leaders of technological companies, you had a much more international perspective. You visited Europe and the United States. You listened closely. You took ideas when it was appropriate to do so. Do you think that the direction of NEC was different from that of other technical companies in Japan, by taking an international perspective?

Kobayashi: My feeling is that we shouldn't fight face-to-face against the world's giants. That is suicide. Instead, we have to look into the niche market, develop a completely new market between two giants, and create a new market in the world. For example, the satellite communications market had not existed before NEC developed it. In this way we can survive. Many other Japanese companies follow the giants. We have always tried to survive in the new market. The overseas market was one such new market. During the 1960s we stepped into international competition. Of course we have many competitors in Japan. But we're always looking to the worldwide system.

Aspray: That's a post-World War II philosophy because the business had been almost exclusively within Japan before the war.

Kobayashi: Right. Later I worked out a six-point overseas strategy. One of the six points says, "If the domestic market is bad, go international." This does not make sense. Overseas markets should be created intentionally.

Aspray: In order to pursue C&C, you had to build up a big research program within NEC, and before the war I don't think there was much of a research mechanism within the company. Yet after the war, a very effective one was built up. What was your strategy for building up a research operation within the company?

Kobayashi: At the end of the war we had a big research laboratory located in a Tokyo suburb. Many scientists and engineers were there. I suggested to the company board that such researchers should be dispersed companywide. This suggestion was accepted by the management, and in May 1949 the research laboratory was disassembled. The situation later changed and in July 1953 a new research laboratory was reopened.

This new laboratory was intended for both basic and applied research. Under this new structure, NEC planned to develop its own breakthroughs as well as follow more advanced technology of the Western world. Earlier dispersion of research scientists into our engineering and manufacturing sectors promoted research-minded operations within the company.

Now, I'll pick one example of our research engineers, Dr. Uenohara. After he graduated from the university, he elected to stay in the university research group. But he was not satisfied with the atmosphere there in the Japanese university. He gave up his position there and went to the United States to work at Bell Labs. In this instance, there is no international border in the area of technology. Technology should be common property of the peoples of the world. No national borders. Years later, I tried to persuade him to come back to Japan. [laughter] Many times. But he was very much satisfied there with his work at Bell Labs in electron devices. Fortunately I was finally successful in persuading him to come back to Japan. We needed him at NEC.

Every engineer and scientist wishes to have his own idea being developed. They need a special environment, for example, some sort of laboratory in order to be satisfied. They don't much care about the results of such technology—to build up new technology from the work of scientists and researchers. I'm not a research man. I didn't intend to limit our technological activity by national borders. If any company asks us to give our technology, to turn it over to them in return for satisfactory compensation, that is acceptable to us. Technology is a commodity. The world community should enjoy it.

Aspray: I see.

Kobayashi: The only problem is: how much is the price of such a commodity. [chuckling] That is the only trouble. Twenty years ago there was no such generosity in treating technology. But now I think it is perhaps coming to be. New technology was developed at NEC before the war, but not in great quantity because the company was too small. We were thus not part of the research and development community. Even then, though, the engineering people had already organized some kind of research and development. So we still thank all the leaders of NEC for the tradition they had built up. This was a nice tradition for NEC. Today in Japan, recognizing NEC as a high-technology company, graduates from engineering universities choose NEC each year as the highest-rated company in which they want to find their jobs.

Aspray: I understand that you were looking to the international market, but you were nevertheless doing business as a Japanese company. In what ways was it an advantage, and in what ways was it a disadvantage, to be trying to do business in the world from Japan? What effect did government regulations, the education of your engineers, protective markets, or other things have?

Kobayashi: I think that there is some misunderstanding for you about this matter. We are not limiting our operations to Japan. We have many NEC companies in the United States, in England, in Germany, mainland China, everywhere. They need to work as a community citizen of that particular country. For instance, NEC-America should do its own marketing, its own development, its own research program, which is suited to or desired by the United States, not thinking of the situation in Japan. NEC headquarters in Japan should give consideration to, for example, each country's national policy. Sometimes

NEC companies of the world compete with each other, or even with parent NEC. I don't mind that. It's very difficult to limit their operations.

We have a big NEC company in Brazil. The reason is very simple. NEC started to support the communications industry in Brazil some twenty years ago. At first we had to transfer our technology and build up the operations there. But after it was going, we expected them to build up their own position in that country. Even if parent NEC company sometimes competes with NEC-Brazil, I don't mind. I leave it to be a very free situation. Sometimes some friction occurs. I leave it for them to solve. We must remember that NEC was born in the globalization process of Western Electric of the United States. Initially they supported us in multiple ways. Year by year NEC grew up to what it is today. Now NEC has grown up to be a world-class high-tech company. Likewise, our local companies might become large enough in their countries.

Aspray: I see. That's very unusual.

Kobayashi: In Australia, we have two or three companies. They are all operated by the Australians, not the people in Tokyo. They are seeking some market outside Australia. I think they have been contributing to the international payment position of that country.

Our basic philosophy for the international operations has been that the final product has to be manufactured at the nearest place to the marketplace. For that purpose NEC established many local companies. They are not for NEC's sake only, but primarily for the local communities. In the developing countries, NEC established these companies partly because of the cheap labor, but in most cases because of the strong demand from the local government, implementing local production projects.

Aspray: There are certain technologies, for example semiconductor technologies, that are very expensive to develop. In order to be able to design those new semiconductor technologies, you put in a very large capital investment at the front end. So you have to have world markets rather than just national markets to be able to make a profit. In some sense you can't compete with one another in those kinds of businesses. It's just too expensive. What is the company's strategy in those areas?

Kobayashi: There are large amounts of investments for semiconductor operations at NEC, both abroad and in Japan. As you know, the semiconductor is a world commodity; size is small and functions are standardized to some extent. Plants and circuitry design centers dispersed worldwide make it possible to produce semiconductor devices at places near customers and in accordance with particular local needs. We have such plants in Asia, North and South America, and Europe.

In the manufacturing process, semifinished products made at plant A could be transferred to plant B in another country for fabrication. This type of world manufacturing network started full-scale operation during the 1980s.

Aspray: Let me ask you one last question. These gentlemen[*] sitting in here with you have clearly risen in the company because of their talent. How did you identify talent and develop it within the company?

Kobayashi: [chuckling] This is one big difference in the idea of employment between Japan and the United States. Usually Japanese people work in a company their entire working lives. NEC is not forced to offer lifetime employment. Employees are free to leave NEC and find a new position in another organization. We therefore need to make the employee feel that NEC is his or her best choice. The company must make him or her think "only here, no other place. This is the best place to work." When employees feel dissatisfied with the company, they might leave; we cannot stop them. Therefore, if such comments are common among the employees, that means that the decisions or strategies of the top management are somehow not right. But most of our employees do not seem to want to change their company. I think this is because our human resource management system has been effective to some extent.

Presently, twice a year, in July and December, all the employees are considered for their job changes within NEC, and eligible persons are transferred to other jobs. Each employee's own desires are taken into account by the management in assigning a new job.

---

[*]Several other key employees of NEC were present at the interview and occasionally helped Mr. Kobayashi with expressing his answers in English. These individuals (all of whom are identified as "Staff") included: Mr. Yasukuni Kotaka (Vice President, Corporate Staff for Engineering), Dr. Michiyuki Uenohara (Executive Advisor), and Mr. Kiyoshi Yamauchi (Chief Specialist, Corporate Communications).

Aspray: Yes.

Kobayashi: I think for me NEC is my life. I'm satisfied. But my long years of service with NEC were not directly connected with income. Even though salaries are not what they are in the United States or Europe, I don't like to appear that I'm against NEC. That is just a feeling. I am afraid it would be difficult for non-Japanese to understand this sort of feeling. Many times this question was raised to me, especially in the United States. It's not a theory, it's a feeling. We need to let all employees understand NEC correctly.

Aspray: I notice you've written much more than most chairmen of major companies. Is this your way of passing on your message to your employees?

Kobayashi: Yes, I'm thinking like that. [chuckling] But I don't know whether our people like that. Once I set forth "Ten Pointers for Executives." The first point is, "Make a picture of your thoughts. Maps and sketches will provide guidance for attaining your next objectives." Writing a book also serves the same purpose. It helps clarify my own strategies before the eyes of the NEC people and also invites other people to evaluate my own management direction. It is very helpful to me.

Aspray: If you recognized a young person as having great potential talent, how would you develop that talent in the company? Would you move them around systematically to a set of different jobs? What would your strategy be for development?

Kobayashi: As a formal job rotation system, NEC has a regular personnel change program. If I remember correctly, this system was introduced at my suggestion. To further elaborate this system, a skills inventory system was inaugurated in April 1965. Under this system, each employee must submit once a year his or her special talent which may be useful to company operations as well as his or her desire as to whether he or she wants to stay with the present job or move to another function area. In addition, we at NEC have maintained a large amount of investments in human resource development programs. NEC employees of overseas local companies are also eligible for those programs. Our belief is that "our human assets are the most important assets of NEC."

Chapter 7
Kazuhiko Nishi

About Ascii

Kazuhiko Nishi founded the Ascii Corporation in Tokyo during 1977 after completing his first and only year of college at Waseda University. Nishi and his two partners launched their small publishing company with thirty thousand dollars Nishi earned from a part-time job and three hundred thousand dollars he borrowed from his father. Ascii published magazines and books about personal computers in its early months and then began publishing computer software soon thereafter.

A year after its founding, Ascii entered into its first joint venture when Nishi reached an agreement to represent Microsoft, the American software company, in Japan. The partnership proved mutually beneficial. Microsoft gained a vital link to the Asian software market through which it could market its products and Ascii's sales grew more than nineteenfold. During his dual tenure at Ascii and Microsoft, Nishi developed the first laptop computer, which in the United States was marketed by Radio Shack as the TRS 80 Model 100. Ascii continued its affiliation with Microsoft until 1986, when Nishi chose to devote his full attention to Ascii.

Once established as a leading producer of personal computer software and publisher of computer magazines, Ascii entered the field of semiconductors. In doing so, it established manufacturing partnerships with Hitachi, NEC, Hewlett Packard, Epson, Matsushita, and Fujitsu. Ascii continues to diversify and expand into new markets. Profits from the Ascii-designed Nintendo joystick were invested into emerging American companies such as Chips & Technologies (producers of chips for consumer electronics and telecommunications) and Ephonics (manufacturer of a relational database). Telecomputing and audio/video/movie distribution are also among Ascii's more recent business areas.

The company went public six years ago. Currently, Ascii is the largest producer of personal computer software and the largest publisher of computer magazines in Japan.

Kazuhiko Nishi

Place: Tokyo, Japan

Date: February 16, 1993

Aspray: Could you tell me in what ways a technical background is important in managing the kind of business that you run here? How important is it to managing software, semiconductor design, and publishing?

Nishi: Technology skills?

Aspray: What kinds of skills are important to senior management?

Nishi: The basics of electronics are important. If a person is in software, he has to understand the basics about software engineering. If a person is in semiconductors, he has to understand all the technologies concerning engineering, and also physics. But most important, I think one needs a general perspective of where our target is ultimately located in the electronics field. And then an ability to create a road map from where we are now to where we want to be tomorrow. We are proceeding in fields where there is no road. So what is important is the talent to write the road on the map where there is no road.

Aspray: Now that the management operations are getting to be fairly large to what degree do you leave technical decisions to others in the company?

Nishi: Basically, we sit down with the management and agree on a target or a budget. Once we've agreed, the rest is just their responsibility. Once every two weeks, we sit down and have a review meeting. But at many review meetings we just listen to the people and point out if there are some mistakes being made, and we say, "This is a mistake."

Aspray: Are there major technical decisions to be made from time to time, not of the character of "Should we enter this market or not?" but rather of the character "Do we use this tool rather than that tool? Do we think that this technology is too hard to develop at this time?"

Nishi: That's all up to the division management.

Aspray: I see.

Nishi: My basic methodology of running the business is to make fifty people comprising one group or one profit center run independently.

Aspray: I see. So, in a sense, there is decentralization in your operation.

Nishi: Yes. I have more emphasis on profit and sales. Telling the people the best way to make the best profit is creative. You need originality for that. We tell our people not to be biased towards the technology. Technology is a tool; it is not our objective. Our objective is a product to sell.

Aspray: Does that mean that you don't engage in so-called basic research within the company?

Nishi: No. Not at all.

Aspray: Not at all. I can understand that in the case of your publishing operations and in the case of your software, perhaps. It's a little less clear to me what the situation is in your semiconductor design.

Nishi: Yes. We have subsidiaries and joint ventures who receive special government funds for research and development. We use such public money for pure research. Then we share what we learn with other people.

Aspray: With the other partners?

Nishi: Yes.

Aspray: What is the advantage of this to your company?

Nishi: These pure research expenditures are not short-term business outlays. They are long-term investments. From common sense of ordinary business activities, it doesn't pay off.

Aspray: I can see it from that direction, but I was wondering whether it was even a good thing to be in at all. I mean, you're looking to the short term with your business strategy, primarily.

Nishi: Primarily, yes.

Aspray: Why enter into these kinds of long-term operations?

Nishi: We have discovered that if, from the center, you focus on short-term, fifty people in a group, twenty to thirty million dollars per group, profitable businesses, your operation does not grow big. It's very hard to grow big. Creating everything from scratch really builds up unique technologies and unique products. It is the only way to build up a 100 to 200 million-dollar business and make a good profit.

Aspray: I noticed from the materials I've reviewed that you have business relations with some very large electronics firms in Japan, such as NEC and Fujitsu. Can you tell me something about the nature of those relationships?

Nishi: We have a lot of joint development, and we have invested a lot in U.S. start-up companies to produce specific products that are very hard to find in Japan. In America there is a lot of engineers' mobility. Because of this mobility, there is a lot of cross-fertilization. In Japan this is impossible. These synergies of technological people make up for the development time lag. By investing in this company, and by making these products available to large companies such as NEC or Fujitsu, who have never thought about the kind of product we are engaged in, we develop successful joint operations. We are somewhat of an intermediary between U.S. ventures and these large Japanese companies.

Aspray: I see.

Nishi: Through these transitions and by serving as an intermediary and seeing how new technology is being developed, we have been accumulating the methodology of how we are going to get this kind of development to happen inside the company. That's what we have been doing.

Aspray: Besides basic research, there are at least two ways that you might gain from a company like Fujitsu or NEC. One of them is that you could use their large manufacturing capacities. If you do some semiconductor designs, they can do the manufacturing. Is that the case with some of your work?

Nishi: Sometimes we do the technical invention, but the product is usually designed in America. And we invest in that company.

Then we retain Japanese distribution rights to anything being sold in Japan that is being manufactured by a company like Fujitsu or NEC. Our partner buys in America. We buy back from them and sell to Japan. So, the partners and myself are most happy with that kind of arrangement. It has been productive.

Aspray: I don't know the Japanese industry very well, but my recollection is that NEC is the largest personal computer supplier in the country. Do you take advantage of that in developing software products?

Nishi: Yes, a lot. Because we understand their market, and they provide us with whatever marketing information is necessary. So we have easy access to the market because we have no competing hardware businesses with them.

Aspray: That's something that they require in their business partner, I assume?

Nishi: Yes.

Aspray: My guess is that a big traditional company doesn't understand the personal computer business nearly so well as a small market-driven company like yours. Is that a fair statement?

Nishi: No. The market is changing very rapidly. So understanding the market is one thing. But the more important thing is having a vision about what is going to happen. With that vision and with what is happening in the market, you have to come up with the next plan of action. With that vision, the company should create an appropriate road map. That's more important. For the last ten to fifteen years I have seen everybody who said "I have a vision," or "I have a great understanding of the market," all fail. Many people make mistakes. The mistakes always happen at the prime of success.

Aspray: Your company does a variety of different things: publishing, semiconductors, and software. Could you tell me how these relate to one another? That will give me an impression of your conception of how you put a company together.

Nishi: First of all, I started the company by publishing magazines. Because I didn't have any money when we started, this was the only thing I could do. Also, no computers existed on the market. So I wrote articles about what kind of computer would come. It was really a type of propaganda. Then we published books, namely translations of American technology books. Then we published software—a lot of software. We became an agent of

Microsoft and imported Microsoft software. Then I discovered what set the boundaries for software. The ceiling is set by semiconductors. So if you can do something creative with semiconductors, we can make the best use of them by combining them with the right software. That's so-called synergy. So we are getting into the semiconductor business.

Aspray: I see.

Nishi: We are also in the telecomputing business.

Aspray: You mean services?

Nishi: No. We sell different types of computer databases. We feel that's a style of electronic publishing. We are also in the movie distribution business.

Aspray: That seems far removed from these other companies and products.

Nishi: The reason why I wanted to be in this business is to be involved in movie production. Today people watch movies in just one way. You turn on the "play" button, and just watch for 150 minutes. When digital audio/digital video is available on computers, somebody is going to think about making interactive movies. I would like to be in that business. So we want to be engaged in the movie industry.

Aspray: I understand the rationale very well, but that seems to me to have a rather long-term payoff.

Nishi: Yes. That's the reason why we just want to pay off running the businesses by distributing movies and building up relationships with other producers or other studios.

Aspray: I see.

Nishi: Publishing, software, semiconductors, movies, and telecomputing comprise our business portfolio. There's enough reason in my heart to justify why we got in to those markets. But we don't disclose our motives. The more people hear that, the more they will ask: "Is he crazy?" But then, nobody really touches this business. Many of the businesses we are in don't produce money. If it breaks even, then the business continues.

Aspray: It seems to me that to be in such a wide range of operations, you need specialists that have lots of very different kinds of technical skills.

Nishi: Yes.

Aspray: What do you do to get that kind of synergy? It's not likely you'll have someone with both the skills of a software designer and a semiconductor designer. How do you manage to bring those together?

Nishi: We haven't come to the point where there is synergy—yet. But what we have now are eight or ten independent groups, independent business units, running independently and each making a profit. So our short-term goal is to have somebody conduct the synergy.

Aspray: Yes. In the 1970s and 1980s, there was a great deal of talk about programming methodologies or various kinds of design methodologies for software. Do you adhere to any of those in trying to produce software effectively?

Nishi: Yes. Methodology is one thing. But it doesn't really make twofold or threefold improvement. Whereas personal differences make improvement. So our approach is to put the better programmer in charge.

Aspray: When you're writing software, what is the typical size of a software writing unit? How many people would be involved?

Nishi: Five.

Aspray: So it's a small enough group that the individual really matters.

Nishi: Yes. Our main product is our computer software. It is not large software. It does not involve tens of thousands of lines of code.

Aspray: I would guess that concerns about reliability and quality are not as important as they are in some big databases for use on mainframes, as in the past.

Nishi: No. We feel once it is known that I have developed a product, then such concerns are part of its maintenance costs. It's like receiving telephone calls about what's wrong with this product. That is an enormous task. So we have an intercompany quality assurance group, and they check the product in advance. They operate the finished product like a customer and see that it really works. Without the approval of that group, a product division cannot launch a product.

Aspray: Do you distribute most of your products directly so that you have personal information from your users?

Nishi: We distribute our products through multiple channels. Books are distributed through bookstores. Computer games are sold through toy stores. Personal computer software is routed through personal computer stores, value-added resales, and manufacturers. They sell it together with hardware. Semiconductor chips are distributed through manufacturers. We deal directly and do things by telecomputing. We use credit cards. We have multiple channels of distribution.

Aspray: To what degree do, say, your software people need to know about the market and the users' needs? When you are looking to hire a new set of programmers, are you looking for someone who has very strong programming skills, or someone who has a sense for a particular user community, or both? What is it that you are searching for?

Nishi: We look for engineers with different strengths. There are different kinds. We seek somebody who has extraordinary talent in one category, and we forgive that person his weaknesses in other categories. Large companies in Japan set minimum requirements for engineers. They have to have at least 60 points in every category. Some require 100 marks in all areas. As I say, I forgive weaknesses in some areas as long as the engineer scores 90 or 100 in a particular field. I would forgive you even if you scored 10 or even 5 or 0. That's fine with me. So many of the people we hire have dropped out from large companies.

Aspray: That's interesting. I don't know what the situation is, but I thought I'd understood that these days it's difficult finding very strong talent in the software business in Japan.

Nishi: That's correct.

Aspray: You are saying that you have the opportunity to provide a different kind of work environment to these engineers—one that appreciates different sets of skills. So you can attract people who may be very good for your company, but might not be good in a more traditional company?

Nishi: The way we recruit the good types of people is this: We hire a lot of part-time employees and university kids. After a few years when we have invested enough time with that person to understand him or her, if we feel this is a good person to work with, we hire that person. It's mainly not our first-class graduates. It is mainly a person who has already established some different relationship with us in part-time work or a summer job.

Aspray: It's not an old enough company to have had to face this too very much, but what kinds of continuing education do you feel that you need to give to your employees? This is such a rapidly changing field. Do you have formal programs for continuing education?

Nishi: Continuing education, we don't have. We don't have any formal programs. But once a year we do a survey of individual preferences. We find out the place where you want to work; if you would like a different kind of task to perform; if you want to switch jobs, and what fields interest you. Based on that information, we make job rotations.

Aspray: Do you find there is much value in formal university courses or short courses? If one of your employees said, "I have a strength in this area but I really need to learn, say, C++, can I go off to take a course in this?"

Nishi: They don't have to go to college to learn C++. But there are some particular subjects that require college, for example, business school for general administration and management. Otherwise, they can learn on the job, or off the job by themselves.

Aspray: Do you send off some of your employees to get special business and management training?

Nishi: We have sent our engineers to Ph.D. programs.

Aspray: Do they continue to work for the company?

Nishi: No, they've just gone.

Aspray: They're just gone for a period of time?

Nishi: One person was gone for almost seven years and just got a Ph.D.

Aspray: I see.

Nishi: He is now gone.

Aspray: So it wasn't a payback to your company. That's a risk.

Nishi: We have friends in the industry who understand us. In that sense we understood that risk from the beginning. I mean, his specialty was not what Ascii needs today. So with a consensus, we let him go.

Aspray: I see. Why is it you think that it's valuable to send some of your engineers off for this kind of training? What do they learn from a formal program in business?

Nishi: Sending one person to the Ph.D. program and having that person come back to the company once in a while and talk about what he's really studying in his Ph.D. research that is different from the operations of the company. Everybody has a clear understanding about the differences in getting their paper written and making a product.

Aspray: What is the typical lifetime for one of your products?

Nishi: There are many different kinds of products. The shortest lifetime of a product is three months.

Aspray: What would be an example of that?

Nishi: Software packages. An entertainment games software package that doesn't sell. If it's going to be a big seller, it lasts two to three years. In the case of semiconductors, if the same product is selling well, by changing here and there and updating it's going to last three, four, five years.

Aspray: But I take it that even your longest-term products are fairly short by general business standards, not computer business, but all business standards.

Nishi: Yes. Somebody who is selling gasoline, yes. Exactly. He's selling the same product for over a hundred years.

Aspray: Right. So that means that you must be putting most of your future development into new products rather than improvements to or enhancements of existing products?

Nishi: By enhancing existing products and changing the division on the product, the product changes as well. It is like publishing books or magazines. Changing the different divisions every month varies the product. Magazines are not really a single-month product. A magazine is a continuation. The product is a subscription.

Aspray: Yes. In fact, there's quite a security once you have a readership built up. You can maintain that if you do your job reasonably.

Nishi: Yes. I think so.

Aspray: It seems to me that may be the safest part of your business in some ways.

Nishi: Yes, it's the most secure income.

Aspray: Nonetheless, there's a great deal of fluctuation in your line of business. Change comes more rapidly than in most traditional businesses. How do you manage that? What kinds of rules do you work by to live in this fast-paced environment? Are there special lessons you can share with us?

Nishi: We just keep going.

Aspray: Just keep going?

Nishi: Well, we have never touched selling gasoline or selling shoestrings.

Aspray: Yes, I understand. But you're unusually successful in a field where there are many, many failures of small companies.

You've grown, you have dynamic products, and so on. So you must be doing something right that many people don't do right.

Nishi: Well, my recognition of success is a little different from your view. We think we're not so successful. We think we have just survived. Since this business has a fairly large profit margin, it was possible to manage the company without careful management. So big money comes in, and you just grab the money, and spend it. The company has kept going on that basis. It's like a mom-and-pop operation. Because of the profitability, because of the size of the market, we have been managing the company on that basis. In that sense we're lucky. But my premonition is that since the market is not growing that fast our management strategy cannot last long.

Aspray: I see. So the flag that change has to occur is not because you've grown to a certain size that you now have to manage the size of your operations. When you're a fifty-person shop, it's a lot easier than when you're a thousand-person shop. Rather it's that the market has stopped growing at the very fast pace.

Nishi: Yes. And maybe we have learned our management style of business by small groups without really digging deep. Expanding the business from magazines to books, software, business software, semiconductors, telecomputing, and movies continued our tradition that business had to be managed in small groups. The key then is who decides which directions to go. That, plus deciding what size the company should be. If all our sales and subsidiaries are going to be over $500 million, then we need fairly extensive managerial attention.

My experience is that budget is a key in giving people the feeling of achievement in business and engineering. We'll let this fifty-people group write the budget themselves. Corporate management, that group made up of general managers and division managers, sits down and really discusses the strategies and decides on the road map and the budget. Once most of them agree to that budget or road map, they delegate the responsibility to operate to the smaller group and let the corporate manager keep them on track. As long as they clear the goals—such as quarterly goals and monthly goals—then we don't really touch them. Delegating authority and achieving goals are really important.

Aspray: Are you building up some sort of central set of business skills that can be used as a resource for all of these different divisions?

Nishi: Yes.

Aspray: What are you consolidating into that business?

Nishi: One, finance. Two, general administration procurement. Third is personnel achievement in management. Fourth is legal expertise: Understanding the contract, and the patent. We are also developing public relations and corporate information systems where each different profit and loss is all consolidated electronically on the same database.

Aspray: It's into these kinds of positions that you want particularly well-trained, hard-nosed business people?

Nishi: Yes. We feel that only somebody who has experience in a large company or a bank or in a credit department can be applicable for these responsibilities.

Aspray: Is it hard attracting people with those kinds of skills to a company that has your profile?

Nishi: Very hard.

Aspray: What do you do to try to resolve that problem?

Nishi: We carefully look around for the person approaching retirement—perhaps five years before retirement. So sixty years old is ordinarily the retirement age. So we approach people at the age of fifty years old when the company offers him a ten-year retirement package. We match this bonus, and then we offer to pick up the person. At fifty years old a person can work fifteen more years. They don't want to retire at the age of sixty. They want to work up to sixty-five or seventy. We go to them and say that if you join Ascii now, you can work fifteen or twenty more years. So these people come to us with their skills. As a result, our managers and executives at corporate headquarters include an ex-CPA, an ex-statistician, a former MITI official, and so on. We are requesting the Industrial Bank of Japan to send us three or four more people.

Aspray: One problem that growing start-up companies face is that people don't grow as fast as the company does. How do you deal with that problem?

Nishi: This has not been a very big problem. However, some of our very senior managers, many of whom started with me from scratch, are facing their personal boundaries. Some cannot grow up as the organization grows. We need to give that person an escape pass. We cannot give him more responsibility because he feels that he is beyond his capability already. We give him a more relaxed job, like a place in some subsidiary. But to the really talented young people, we give even more responsibility and accelerate promotions.

Aspray: They are two back sides of the same issue. Have you moved some people, some talented young people, rapidly through the company?

Nishi: Yes, any people who joined the company at the age of twenty-two, we treat them very equally for the first five years. Then there is going to be a difference. The Japanese really care about the position of their classmates.

Aspray: It seems that that's a very sensitive issue here.

Nishi: It is a very sensitive issue.

Aspray: One wouldn't face such an issue in the United States. Yet within the rigidity of the Japanese work system, you are able to accommodate your needs? You are able to move people at different rates?

Nishi: Two issues are key: responsibilities and salary. By keeping salaries the same, we give the talented people more responsibility. After this person achieves results, then we change their salary.

Aspray: Are salary and benefits important in retaining good talent within your company?

Nishi: Yes. We always say: "Flower in both hands is impossible. Flower in only one hand." If you give responsibility and salary at one time, both at the same time, the person feels that they become king. That is going to ruin the person's life. So we really have to be careful about giving responsibility before raising the salary.

Aspray: I notice that you have some subsidiaries in several countries. What is your strategy for growing beyond the boundaries of Japan?

Nishi: We have some operations in the United States. We invest in companies. In some of the companies we own the majority, in some we own just a few percent. For the time being our primary business focus is Japan.

Aspray: I notice your publications are only in Japanese at the moment. I don't know the situation with your software products, but do you feel that you need access to more than a Japanese market to recover your costs to get the appropriate kinds of earnings on them?

Nishi: My business pays off within Japan.

Aspray: Within Japan?

Nishi: Anything overseas is just extra. My basic belief which has evolved is that the Japanese market is one tenth of the worldwide market. For example, Microsoft is a large-scale company. It is about a $4 billion company. Our company is about $400 to 500 million. Its value is one tenth of Microsoft. So if we go after the worldwide software market, then we have to be ten times larger. But the issue is, can we manage such a worldwide organization? That's a different issue. I choose to diversify horizontally in Japan and deal exclusively in the Japanese market. Microsoft is very weak in publishing. They publish some books, but that's really it. So I see in my businesses, two categories. One is content-oriented businesses. Publishing, entertainment software, audio-video software, telecomputing services are content businesses. The other is functionally-oriented business that includes database software and semiconductors. We have to go on worrying about our semiconductors because our equipment is all original technology. But the software business is America; American software companies are five to ten years in advance of Japanese businesses. So it's tough to compete over there. Publishing is a very culturally oriented domestic business. So there is no point in expanding to America in that area.

Aspray: That was to be my next question. In what ways are the kinds of businesses you do international or national? Would a product sell just as well in England or the United States if it sells in Japan?

Nishi: In the publishing business, the target is Japan. Entertainment software, we think we can sell worldwide. Business software, we sell only in Japan. Semiconductors are a worldwide business.

Aspray: You mentioned in one of your earlier answers the issue of patents. What role do they play in a company like this?

Nishi: Patents are really key for protecting us against competitors producing the same product. It really protects the lead position of the product. We feel that protecting our lead is one of our key strategies in semiconductors, but not in software.

Aspray: You have apparently grown to a position where you're spending a great deal of your time on management issues rather than in publishing decisions.

Nishi: Yes.

Aspray: Can you tell me about your adjustments to this? What in your formal education, and what in your experience, was useful for you to do this? What lessons have you learned over time? What mistakes have you learned from? What did you do right from the beginning?

Nishi: Basically, I didn't go to college. I only went there one year. At the end of that first year I started my company, Ascii. So I had no college education. I learned business on the job, starting the company from scratch. At that time my partner was the president, and I was the executive vice president running product development. I became president six years ago, and then we went public to get money. We invested the money. Our continuous growth can be attributed to the growth in the market. So even without management we have come to this point.

My premonition is that we are at the point where we can get organized. If we can get organized, we can be a billion-dollar company. If we get organized, we are going to win. It's just a matter of time. That's my basic understanding. So I am making an effort to systematize this fifty-people business organization, including research and development. That's one thing.

The other thing is how I spend my time. I compartmentalize my time. One part is spent on management in businesses especially. Another one third of my time is for my personal research and development, where I do my own programs by myself. Plus I teach at the university.

Aspray: What do you teach?

Nishi: I teach things for the media, systems engineering. I teach two hours a week. But because I am busy, I'm requesting four hours every two weeks. Then I can travel. I have my own staff for my personal research and development. It's like brain athletics.

Aspray: I'm sorry?

Nishi: Brain athletics. Like you train by running in tennis. Your brain has to really do something like that. One, because management is something that you don't really do. You just say "yes" or "no." You just show the road map. For example, when I am working as a manager, I have only a red pen. When I am working in my personal research and development, I have no red pens, just a pencil for drawing the pictures. One third is my personal life, which is my family—my parents. I used to bundle my personal development together with my personal life. At that time I was very unproductive. By having stronger mental concentration, you can improve the productivity by two times or even three times easily. Then using the rest of that time for something different. We have discovered it is very important for the balanced life.

Aspray: I can see that. Has capital been a problem for the company?

Nishi: It was not a problem in the past; but today it is a problem. Today, a lot of Japanese companies are suffering from inability to get the money from the bank.

Aspray: What is the way of solving that problem? What are your options?

Nishi: The only option is making the product that will create a lot of cash flow and cut down on expenses. Cut expenses to the minimum, increase the cash flow to the maximum. It's a very traditional style of arrangement.

Aspray: Is it possible, though, for example, to take in a bank as a partner? I know one particular software company that's about your size that has recently had financial problems and has decided that the way to handle it is to take in a large bank as a partner, give some seats on their board, and so on.

Nishi: Yes, that is what we do. We are asking the Industrial Bank of Japan to send management to sit on a board seat, and give us a loan, some credit lines. We only invite long-term investment banks, not merchant banks. Ordinary merchant banks cannot cope with the problem we are facing.

Aspray: What would you say differentiates your company from other companies in your business areas in Japan? You have various business areas, and so you may have different competitors in each of them.

Nishi: Yes. In every category our size differentiates us. We are the largest in computer publishing, we are the largest software company in total volume. In semiconductors we are very special. No one in the software industry touches semiconductors. So we maintain our unique positions. Our feeling is that our competitors are not other companies. Our competitor is ourselves. If you don't touch an untouched area, that's a mistake.

Aspray: What other kinds of management lessons or management philosophy can you tell me about for your company? We're looking at various people's management philosophies and trying to understand how to manage a technological business better. What other lessons can we draw from your company that we haven't talked about already?

Nishi: I think a key is a budget system. When we are writing the budget for the next year, we assess the talent and the capacity of each group, and then agree on the sales and the profit goals. That's really the key. If you impose goals that are too high, then the manager is going to end up writing an unrealistic budget and they sort of destroy themselves. If the manager sets easily achievable goals, the worker doesn't really work. And if you look at other groups of the same size making more profit, more sales, then you look foolish. So I typically assess the next year's goal at ten percent higher than last year.

I then give that division appropriate financial, technical, personnel, and sales support. We incorporate lots of work stations but connect them in a local network. Ascii is a united company, a united division. That is my concept. I have learned that if you are going to diversify horizontally, the key is to create a system that will react in real time to the changes of the market. You cannot make every decision by yourself. You really have to delegate. But what are you going to delegate? My style is delegating execution, but I don't delegate budget planning. That's consensus between that general manager and myself. As long as I don't agree with a division's budget goals, that division doesn't start business. If the employee is not doing good business for three or four years, then I say, change him because he violated the promise for three years. That's the system.

Long-term investment—a five- or six-year development project—is another issue. We create a separate research field and invest the money and run it differently. The key is we can make money not really doing high-tech stuff. The best way to make money is to do the low-tech stuff. We make a better profit margin by selling not books, but magazines. So the company should be divided into one part, the division generator, and the other should be development.

Aspray: I was wondering if you could briefly tell me about your personal history. I know that American audiences, at least, don't know much about you. You went to Waseda University?

Nishi: Yes, that's right.

Aspray: What were you planning on doing? What were you studying there? What were your career plans?

Nishi: I wanted to study robotics. Then the professor assigned me to a robot-computing computer. So my interest is in numerical control.

Aspray: Why did you decide not to pursue that direction?

Nishi: There are a lot of boring undergraduate classes. Instead of going to these classes, i|'s more fun to run the business.

Aspray: How did the business get started in the first place?

Nishi: I wrote a lot of articles on computers. I submitted these articles to a lot of magazines, and a lot of magazine publishers rejected them. So I published them myself. That's the body of the editions for our magazines.

Aspray: I see. And how did you make the decision to become a company? I mean, one can do this as a hobby for a short period of time. At some point it has to become a business decision.

Nishi: I always wanted to be organized. So we assigned a person the responsibility of running it, assigned one person the responsibility of sales, and one person the responsibility of editing—that's organizational diversification.

Aspray: It's always hard to start a new company.

Nishi: It was not that hard.

Aspray: It wasn't that hard? Why was that?

Nishi: The reason is quite easy. There was no personal computing industry sixteen years ago. We were the first maker of computer publications. We had no competition.

Aspray: But you had to find capital at least, you had to learn about doing business, and so on.

Nishi: I accrued the key money, which was thirty thousand dollars, from my part-time job. I went to my father to borrow three hundred thousand dollars.

Aspray: You alluded before to having a partner. What was the division of labor? What were your responsibilities, and what were his?

Nishi: I had two other partners. They left almost a year and a half ago. One was the chairman, and one was assistant vice president. The chairman used to be president. He was running the business since I was running the product creation. The assistant vice president was in charge of publishing. He was running the publications group.

Aspray: So your publications business began to expand over the first few years of the company's operation? How did you decide to move beyond publications and get into other lines of business?

Nishi: I went forward with the publishing business.

Aspray: What were your first moves outside of publishing?

Nishi: Software. After software was semiconductors. Then telecomputing. Then audio-video—movies.

Aspray: I understand that you had a close business relationship with Microsoft.

Nishi: Yes.

Aspray: Could you tell me something about that?

Nishi: From '77 to '85 we were just representing Asian Microsoft, selling their software before it went public in the United States.

Aspray: That terminated at some point in 1986?

Nishi: 1985.

Aspray: Why was that?

Nishi: They wanted to start their own Japanese operation.

Aspray: So they wanted to control?

Nishi: They didn't like us being in a lot of other different businesses. They want a company that just did software.

Aspray: I have read that you were involved in one of the early laptop processors.

Nishi: Yes, the Modem 100.

Aspray: Can you tell me the history of that? How did that come about?

Nishi: We created the product together with the gentleman in Radio Shack who later became president of Microsoft, John Sherry. They really refined the product.

Aspray: There was no such product on the market at that time?

Nishi: No. It was 1982. Ten years ago.

Aspray: Can you tell me about your relations with Radio Shack? How did this come about? Who had the idea for it?

Nishi: We designed a product together with a company called Kyocera. It was one of the top companies in ceramics. We created our product together with our chairman, Kozumo Morii. His engineers worked hard to produce a product and went to Radio Shack for possible distribution.

Aspray: Is it because they had such a strong distribution network?

Nishi: Yes.

Aspray: Did you distribute the product in Japan?

Nishi: No. Japan was being distributed by NEC, and Europe was distributed by Olivetti. It was a very successful product.

Aspray: Did you have a personal involvement in the product design?

Nishi: Yes.

Aspray: Looking back over the history of the company, what would you say were the few key events?

Nishi: There are two key events: Starting the company and the present. My feeling is that you always focus on now. An accumulation of "now" is really what you are today. The trees in Southeast Asia, because of the temperatures, are always very flat. They have no rings because the temperature's always flat. In Japan it's hot in summer, cold in winter so the trees have rings. So the company itself has a lot of good times and bad times. The accumulation of good times and bad times made the company very strong. In that sense the first step was to change its relationship with Microsoft. The second was the departure of the ex-partners. The third critical time is probably today. We need to decide how the company's going to get financed for the next few years to come.

Aspray: What do you look back on as your most successful products? It may not be that they were most successful financially. They may have helped the company move in a direction you wanted it to move.

Nishi: Publishing magazines, and our joystick, we developed for a whole family of computers.

Aspray: I see. It was in the development of your company?

Nishi: Yes.

Aspray: I didn't know that Ascii developed the joystick.

Nishi: Nintendo's joystick is our product. It has sold almost two million units in the United States for Nintendo. So the Nintendo joystick is ours. We invested the profit into the company called Chips & Technologies, and the company called Ephonics, a relational database. We had very high visibility working with Microsoft, but we really didn't make a big profit out of that business. They made a lot of profit. So it was very lucky that we had experience of getting into the software business through Microsoft activities. Our own software is really the software creating a lot of profit.

Aspray: Very good. Are there any other things that you'd like to say to me today?

Nishi: Just about history.

Aspray: Yes. Go ahead.

Nishi: I like to read, I like history. My expertise in college teaching is media systems engineering. The media system's definition is not just a computer, but telephone, radio, television, cassette tape, records, video disk, video cassette, that kind of packaged media and communications—broadcasting media—as I explained.

Aspray: The way that it's defined by the MIT Media Laboratory?

Nishi: In the sense that media is a pipe to convey the content, and we have a big interest in that. But when we look back in the history of this media development, which is my interest, many people ask if we have done this, if we had lots of luck. If the product has been finished on time. But I think that's wrong. We have to go back and review the history by asking "why?" You need to really look back at the history and find out why certain decisions worked. In that sense we find out the true reason, the true factors, for how things have happened. Then we should apply "if" for the future. What *if* we do this? What *if* we do that? Because we do a lot more future stuff, my personal inclination is to go back and study the history. I read many history books about how the personal computer was born and similar topics. But in many history books, about 90 percent of the description is wrong. It doesn't describe the truth. People only write about the bright part of the history.

Aspray: What kinds of things do you think are being left out?

Nishi: The dark part of the history is missing. Mistakes should be included. I personally study the life of Napoleon. I have an original collection of his drawings and writings translated into Japanese. They are very interesting. A lot of Napoleon's biographies contradict each other. Historically, it's fine. But the funny thing is that if you read Napoleon's very close subordinates' diaries, or subordinates' books, writings about Napoleon are all consistent. I am always wondering when reading the history books what is really the truth and what is really the black part.

I started reading my old friend [Bill] Gates's book. I stopped reading after a few pages, because that book is just full of the bright part. Of course many historians have to assess what is really the truth because of these books. It also takes about a hundred years or two hundred years after the death of an important person to really assess that person's actions. I don't really intend to be the person who puts his name and future on the whole of something. I just want to see the history and let readers know what is really the truth. Also I'd like to make my personal assessment about why so-and-so happened and use that information to help me make the best decisions for my life and for my company's life—for my own activities.

I have a huge, ten-page poster covering the past hundred years, which was drawn by Raoul Dufy, of the "Spirit of Electronics." It is my favorite poster and painting. It contains one hundred inventors of the last one hundred years. It was written almost fifty years ago, so it covers a period up to 150 years ago.

By looking back . . . I discovered one very interesting thing: Engineering and technology are really magic. Engineering technology is the methodology of making the impossible possible. It hasn't changed for almost a thousand years. I think it's probably human instinct that we create something new. That's the emotion, or the instinct, that I really would like to make much of. I think that's probably a very important part of the energy of this company.

Aspray: Today I haven't tried to ask any historical questions. It would be possible to try to understand the development and reception of the laptop or the publishing business and such. Some day I think somebody should do that. But I think you're right, that we can't get the perspective yet. We're too close to things.

Nishi: Just simply repeating "what" for the past is not an enjoyable thing. We developed the laptop first. Now we have the best-selling laptop. I have done some important things. I was a member of the MS-DOS development team with IBM. I developed the laptop. But your next steps start when you deny what you have done. To totally deny and destroy what you have done is the beginning of your new activities. By knowing is one thing. By practicing and exercising that is another thing. One time somebody, some newspaper or magazine, gave me an award for the man of the year. I received that, and then came back with that award and put it into the shredder. [chuckling]

Aspray: I see.

Nishi: This is a secret. It was too rude for them.

Aspray: Of course.

Nishi: I was feeling very pained about doing that. I tested my thesis as a practice by doing that. I have successfully diminished my feeling of arrogance. That's really arrogant and senseless. These are two dangers when I say I'm a big shot and that the company is Number 1. That kind of feeling is very dangerous. My opinion about new technologies or innovations centers around the quick action, but also experiencing a business vision and setting it into policy. This is very important. That's my opinion, and I always want to be very sensitive. I always want to be active. That's my goal.

Chapter 8

Takashi Sugiyama
Takashi Yamanaka

About Yokogawa

Yokogawa Electric Corporation (YEC) was founded in Tokyo as an electric-meter research institute on September 1, 1915, by Tamisuke Yokogawa, Ichiro Yokogawa, and Shin Aoki. Their institute was incorporated as Yokogawa Electric Works, Ltd. in 1920.

Yokogawa has long been a leading innovator in the areas of electric measuring instruments and process control. In the 1950s Yokogawa introduced Japan's first electronic tube automatic self-balancing instruments, a digital electronic process-control computer, data processors, and magnetic flowmeters. The 1960s brought Japan's first direct-digital control systems, B/M meter basis weight and water-content measuring instruments, and vortex flowmeters. Among the many innovations of the 1970s and 1980s are Yokogawa's ultrasonic diagnostic system (U-sonic RT), programmable recorders, digital oscilloscopes, and arbitrary waveform generators. Yokogawa has been a frequent recipient of government and private awards for its inventions.

In 1957 the company established Yokogawa Electric Works Inc. in the United States, thus beginning its global expansion. Subsidiaries all over the world such as Yokogawa Electrica do Brasil Industria e Comercio Ltds. (1973), Yokogawa Electric Europe B.V. (1974), Yokogawa Corporation of Asia (1976), Hankuk Yokogawa Electric Co. Ltd. of Korea (1979), Yokogawa Elecrofact B.V. in Holland (1983), and Xiyi Yokogawa Control Systems Corporation in the People's Republic of China (1985) followed thereafter.

Yokogawa has been unusually successful in its joint ventures. In 1963 Hewlett Packard and Yokogawa began a long and fruitful relationship by establishing Yokogawa-Hewlett Packard, Ltd. in Japan, a partnership that continues today to produce computers and electronic measuring instruments for people worldwide in business, industry, science, engineering, health care, and education. Yokogawa's cooperative efforts with General Electric Company (GE) include Yokogawa Medical Systems (YMS), established in 1982 as a leading producer of X-ray, CT, ultrasonic, and MRI equipment. A year later Yokogawa Electric Works, Ltd. merged with Hokushin Electric Works, Ltd. to form Yokogawa Hokushin Electric Corporation. More recent joint ventures include Yokogawa Johnson Controls Company, a producer of measuring and control instruments established in 1989, and Yokogawa Cray ELS, Ltd., the product of a 1991 joint venture with Cray Research Inc.

Presently, Yokogawa has grown into an immense multinational company with over twenty principal subsidiaries around the world. Its diverse product lines include industrial automation systems, industrial sensors, recorders, test and measuring instruments, temperature controllers, meters, analytical instruments, aeronautical and marine products, and building control systems. In recent years at Yokogawa, annual sales exceeded $1.7 billion.

Takashi Sugiyama

Place: Tokyo, Japan

Date: February 18, 1993

I. Written Answers (Prior to Interview)

Question 1 What are the key elements to making your business run successfully?

Answer (a) YMS (GE and YEC Group) has quickly introduced new products such as X-ray, CT, MRI (Magnetic Resonance Imager), and ultrasound equipment—applying current technology and matching it to the Japanese market. There has also been a trend of new instrument introductions into hospitals.

(b) GE and YEC got together to develop new instruments. GE recognized the autonomy of YMS in some areas. YMS emphasizes high reliability and quick development, while GE exhibits the current high-technology-oriented intention.

(c) Mr. Sugita, who was a former medical equipment business division manager at YEC, developed the new market and name of the medical equipment business in Japan. GE's participation in YMS made the expansion of YMS business in the medical in Japan easy and quick.

Question 2 In what ways are a technical background important in managing your company?

Answer (a) Development engineers are able to work without anxiety, because top management whose background is engineering can understand the engineer's standpoint well.
  (b) Customers easily accept technological explanations of the top manager when troubles occur with customers.
  (c) YMS is a technology-oriented company, and if the top manager does not have a technology background, it is not easy for him to decide whether a development project should go or not.
  (d) When the market size is rapidly expanding, as was the case with diagnostic imaging equipment in the 1980s, a new product which includes new technology is very important to expand the company's share.

Question 3 Are there specific experiences in your background that you find valuable in preparing you for senior management?

Answer (a) Improvement of efficiency of product development in Engineering, two examples:
  (1) In the developing process of ultrasound equipment, eight years ago, the sales manager requested a mechanical sector probe, but I decided not to employ a mechanical one but instead to employ an electronically controlled probe; now the mechanical sector probe has disappeared.
  (2) On the way of developing MRI system, there were several kinds of magnetic field strength (0.2T, 0.3T, 0.35T, 1.0T, 1.5T). I agreed and understood well this decision, now GE (YMS)'s market share in the world is Number 1. This strategic success and valuable decision was possible because of my technological background.
  (b) In 1982 I was a senior managing director of Yokogawa Electric Corp. in charge of Engineering, and I moved to YMS as President. At the same time, 333 Yokogawa Electric employees moved to YMS, including sixty engineers. Surprisingly, those engineers joined this new company without anxiety. Their capabilities were better than average level. This shows the importance of the background of top management.

Question 4 What qualifications are needed in your company to become a senior manager and why?

Answer (a) Senior management who are in charge of Engineering, Servicing, Manufacturing, or Marketing have to have engineering background because YMS is developing some equipment applying brand new technology.

(b) In general, an engineer tends to concentrate on his own specialty and not to broaden his view. This type of engineer is not suitable to become a senior manager in the company. A senior manager needs to have a strong curiosity and to be aggressive for many kind of things.

Question 5 How important to your company's livelihood are quality, maintainability, and reliability of products, and what means are used to achieve them?

Answer (a) Since a medical diagnostic instrument directly examines patients, a breakdown of the machine has a sometimes fatal effect, so maintenance should be done quickly and reliably during the off-time of machine (midnight or early morning). Moreover, service stations should be located within the distance of one-hour drive to every hospital, and be open all the time. (Twenty-four hours operating system.)

(b) The most important part of quality is image. The resolution of image determines whether the small heterogeneous matter exists or not, so a YMS product is designed with image quality as the most important item.

Question 6 How much do your employees need to know about the operations and needs of their customers in order to produce successful products?

Answer Recently YMS developed a new mind-setting movement, that is, that our customers are not only medical doctors, but also nurses, technicians, and patients; and all employees, especially engineers and manufacturers, should know the opinion of all of these end customers. YMS collects questionnaires from customers several times a year and reflects the improvement in the next group of new products.

Question 7 What differentiates your company's operation from your competitors in Japan and other countries?

Answer Because YMS is a joint venture with GE, YMS gets more information from the USA than its competitors, including European companies. A recent trend is that the center of medical business and academic study is moving from Europe to the USA. So YMS has advantages over its competitors in this sense.

Question 8 What role does service play in your company's vitality? Is the company's strategy to provide services, provide products, or solve problems (no matter what mix or service and products is required)? What are the most important elements in achieving them?

Answer I already mentioned the importance of service. In a broad sense, service includes improvement of old machines up to the level of recent machines, free of charge. That is the concept of "Continuum" of GE and means the relation with customer is not temporary, but continues for a long time.

Question 9 Are there ways of achieving economic scale in research and production?

Answer The GE medical system division has established a three-pole organization, consisting of the Americas (North and South), Europe (including Eastern Europe and all Russia), and Asia (India, East Asia, China, and Japan). YMS is supporting not only Japan but also the other two poles with YMS products. At the same time YMS imports the products from the other two poles. Thus YMS enjoys its scale market of products and can reduce manufacturing cost. YMS exports constitute 40 percent of total sales, and imports around 15 percent.

Question 10 What commitment does the company make to research and development (R&D)? How are they organized in your company? What are the key management issues associated with R&D?

Answer (a) Total expense of R&D in a year is around 10 percent of product sales. Expense of R&D includes patent royalty paid to GE and other companies.

(b) The choice of research themes and their schedule control are important issues. The former one is decided under the system of worldwide product planning, in the total marketing meeting attended by representatives of all three poles. Schedule control is done by applying the Tollgate system. At each gate, design review is performed.

The key management issue in R&D is not to start new themes, but to stop the running theme. The execution of stopping is difficult in general, however. Decision making and understanding are responsibilities inevitable to senior managers.

Total staff for R&D is two hundred, equal to 12 percent of total employees. Among them research members are around 20 percent.

Question 11 What role do other manufacturers and other industries (e.g., suppliers, assemblers, or service and system companies) play in the operation of your company? How does this affect your business practices?

Answer Since YMS does not have a machine shop, it is totally dependent on outside manufacturers for mechanical components and processing. Many other electric and electronic components are also purchased from outside vendors in the world, according to the global sourcing plan of GE.

The key components such as probes in ultrasound equipment and the X-ray detecting array are manufactured in our company. The total number of vendors associated with YMS is around four hundred.

Question 12 What role does the government play in the operation of your company?

Answer Prior to the new product introduction, a manufacturing company needs to get the license in accordance with PAL (Pharmaceutical Affairs Law of Japan) under MHW (Ministry of Health and Welfare). Other relations with government are the same as an ordinary Japanese company. There is no special strong control or limitation.

Question 13 Are there other kinds of alliances with government, business, or academia that affect the way your company operates?

Answer (a) There are no special alliances with government, business, or academia. YMS keeps good relations with Keio University. Many of YMS's new products have been tested clinically at Keio University. Test results are published at the medical academic society with the name of Keio University.

(b) YMS has a special alliance with GE Medical System Group as a member of the three-pole system. This group exchanges information on competitors' status, technological news, or economical conditions of each geographic area.

Question 14 What kind of qualifications (especially educational) do you seek in your professional employees, and is it difficult to locate and recruit highly qualified people for professional positions?

Answer YMS usually hires people who graduate from the Master course as especially educated. It is difficult to recruit highly qualified people. Occasionally, YMS hires a highly qualified specialist as a consultant. This might be someone who has already retired but who has many precious experiences in his special field.

Question 15 What kind of continuing education do you provide for your employees and is it necessary and effective?

Answer There are about sixty correspondence lecture courses at YMS, and every employee can select any kind of lectures. These courses are provided by professional institutes. Every year around two hundred employees take these courses. The orientation course and T.L.P. (Technology Leadership Program) have just started.

The feature of the educational system in the Japanese company is to educate the whole workforce (including the ordinary workers). On the other hand, the education system in the USA aims at the level of manager or leader.

Question 16 How does raising of capital affect the company's operation?

Answer When a company wants to start new business or to build a new facility, top management would not take an action such as raising capital in general. Top management of many Japanese companies would prefer to operate in the range of its existing capital or rent some funds from its main bank.

Question 17 How free is the company to undertake long-term planning and avoid issues of short-term return in investment?

Answer Our company (YMS) has two kinds of plans. One is a three-year forecast (MRF) and the other is every year's one-year plan. After one year passes, the next three-year plan will be built. The longer plan is usually announced by the top manager at the beginning of a new fiscal year, but the content is not so concrete, but only indicates our long-term target or vision.

In the case of more basic research, it takes longer to become key technology for YMS. YMS sometimes asks the help of GE CR/D or YEC R/D, depending on the content of the research.

Question 18 What is the company's strategy for growth? For example, to make new products, to enter new market niches on a regular basis, to make incremental improvements in products, to increase penetration in market niches in which the company already operates, or other strategies?

Answer In the past ten years, innovative new products have been introduced in the medical diagnostic field. They are ultrasound equipment, X-ray, CT, and MRI. The GE Group (including YMS) could follow this technology change and has introduced new products on a timely basis. After the new product was introduced, model changes or incremental improvements should follow in order to expand market share. As a total, 80 percent of development man-hours are usually spent for incremental improvement of products, 10 percent are spent for new products, and another 10 percent are for special product development requested by salesmen and medical doctors.

Question 19 How do you deal with the rapid changes that occur in your industry?

Answer YMS has a relatively strong marketing staff. Marketing specialists carefully watch the market tendencies, where the market is going and when applicable technology is available. Some specialists attend academic meetings to catch the new information from doctors. (A good example is kidney-stone smashing equipment). We have to pick up some of the new product candidates, and then engineers start development, supported by marketing.

Question 20 What general management lessons can I learn from your company?

Answer You can learn from our company three items:
 (a) Education system in YMS
 The YMS education system is for all employees. YMS wants to educate the whole workforce, including ordinary workers. On the other hand, GE (or other American companies) want to educate management-level employees who will be leaders in some divisions of their companies.
 (b) International mind
 All YMS employees try to have an international mind. In every stage of development, engineering, manufacturing, marketing, and even sales, the staff always thinks about other countries and they try to match the foreign countries' conditions.
 (c) Technology-oriented company
 YMS would like to win against competitors not with sales power or advertisements, but with the product's technological features.

II. Interview

[*Note:* Eiju Matsumoto, curator of the Yokogawa Museum Office, is present at the interview and helps with translation and interpretation.]

Aspray: Why don't we begin with some basic information about your life and career? Can you tell me something about when you were born and what your family was like?

Sugiyama: I was born in Osaka in 1924. I grew up in an engineering environment. My father was a chief engineer at Mitsubishi Electric Company in Kobe. I joined the University of Tokyo, and in 1947 I graduated from the Electrical Engineering Department of Tokyo Imperial University.

Aspray: What was the course of study in electrical engineering at that time? Did it emphasize power or electronics?

Sugiyama: At that time, these departments were quite separate. We learned power electric and wireless communications system, so-called weak electrics.

Aspray: Upon graduation you joined Yokogawa Electric?

Sugiyama: Yes.

Aspray: How did you choose to work for them rather than someone else?

Sugiyama: First, I wanted to stay in Tokyo. I used many instruments at the department of the university. I liked Yokogawa instruments, oscillographs, and some kinds of indicating meters. I wanted to design and manufacture in Yokogawa Electric Works, so I chose Yokogawa Electric Works. That was the number one reason. The second reason was my predecessor's influence. Sometimes I met with Mr. Numazaki. He had already joined Yokogawa Electric Works. He strongly recommended that I join Yokogawa Electric. Then I decided to come.

Aspray: What was your first job with Yokogawa Electric?

Sugiyama: At the time, there was no training course. My section manager gave us a special subject, the designing of an electric low-pass filter circuit.

Aspray: Can you tell me about the various positions you had in the company as the years went by?

Sugiyama: First I joined the research and development group in 1947, and I designed some kind of filtering system, and then I designed many kinds of electrical instruments, measuring instruments. For example, a vacuum tube voltmeter and peak voltmeters, also using vacuum tubes. The voltmeter was very popular and important at the time.

Aspray: These were replacing some older kinds of meters?

Sugiyama: Yes.

Aspray: Had those older ones used vacuum tubes?

Sugiyama: Yes. The first ones only used one tube in the voltmeter. We designed then with two or three tubes. They had a more accurate, wider frequency range. We designed them because vacuum tube prices went down very rapidly.

Aspray: Things must have been very difficult in Japan right after the war.

Sugiyama: Yes.

Aspray: How did it affect your workplace?

Sugiyama: Conditions were not so good. Vacuum tubes were very precious. The working place was very bad and we didn't have air conditioning. Only a stove. But I was young, and it was exciting to devise instruments.

Aspray: Yes. So you designed a series of these instruments. Then what did you do in the company? Did you move into a management position of a small design team?

Sugiyama: Yes. Twelve years after graduation I was appointed as section manager. I had ten engineers.

Aspray: What projects was your team working on?

Sugiyama: Mainly measuring instruments using vacuum tubes. Then we moved to other projects. Around 1955, the digital voltmeter appeared. We started studying this kind of new digital technique. I am very interested in A-D (analog-digital) converter systems. I already had many patents.

Aspray: Very many.

Sugiyama: I found a very precise A-D converter system. I wrote a thesis on it. I sent this thesis to the University of Tokyo, and I received a doctorate degree in 1970.

Aspray: I see. Is that a common way to get a doctorate degree from the university?

Sugiyama: Yes. In Japan, there are two kinds of doctor degrees. One is by taking graduate courses, but there are other ways. In my case I graduated from the ordinary course of study and while working in public I got my doctorate degree by thesis.

Aspray: By the thesis.

Matsumoto: His thesis, "Pulse Width Modulation Aid Convertery," is very famous.

Aspray: Yes. Very important technology.

Sugiyama: I sent this thesis to IEEE. Many letters came to me asking about this thesis. Fortunately, I got a very good team member, Mr. Yamaguchi. He is a very clever guy. He helped me very much.

Aspray: I see that you were then appointed to general manager for Research and Development in 1971.

Sugiyama: Yes, that's right.

Aspray: Had the company had a major activity in research and development before this time? Or was it growing?

Sugiyama: Yokogawa was originally a very technology-oriented company. Many famous engineers were there. Especially Dr. Miyaji Tomota. Dr. Tomota was one of my predecessors, and a former president with an engineering background. He held over four hundred patents. He's very famous. He's also an IEEE Life Fellow. He is eighty-five years old now. We did not have a research group, only a development section in every division. In 1971 Yokogawa started a research and development center.

Aspray: You were the first director?

Sugiyama: Yes, that's right.

Aspray: How many people were working in that group?

Sugiyama: Forty-seven people.

Aspray: How did the work in the new group differ from the way that it had been before? Was there more basic research now?

Sugiyama: I selected some engineers from every division: Twenty people from the control instruments division, and twenty-seven people from the measuring instruments division. The total was forty-seven. Their capability was more basic than other engineers.

Aspray: I see that only one year later you received a prize. Can you tell me about that?

Sugiyama: The Prime Minister Award for invention of pulse modulation technique in 1972.

Aspray: Ah, from your dissertation.

Sugiyama: Yes.

Aspray: I see. This is a very prestigious award.

Sugiyama: I think so.

Aspray: Yes. I see that your career was rising very fast because you were appointed only one year later to the vice president's position. What were your responsibilities as vice president?

Sugiyama: I was in charge of research and development and in charge of the total engineering group.

Aspray: How large would this be? How many people?

Sugiyama: It was around six hundred engineers.

Aspray: Oh, quite large. I understand that you must have had very good management skills as well as being an accomplished technical person. Did you ever take any training as a manager or did you learn on the job?

Sugiyama: I don't know, but it's maybe a natural talent.

Aspray: I see.

Sugiyama: Excuse me. I would like to clarify one issue. In those days, Yokogawa didn't have a matrix system. There were many divisions. These divisions had the engineering group, a manufacturing group, and the sales group. It's a functional corporate group with marketing and R&D. This started with forty-seven members, and I controlled this engineering group.

Aspray: I see, all across that whole engineering group.

Sugiyama: Indirectly controlled.

Aspray: Yes. I see. What were the great challenges at that time for you as a manager of R&D in engineering? What were the biggest problems you had to face and solve?

Sugiyama: This matrix system is good, but actually it was very difficult, because the division's control was very strong and engineers did not understand what I thought. Number two was technology transfer from corporate R&D to engineering.

Aspray: It was difficult?

Sugiyama: Yes.

Aspray: What kinds of methods, mechanisms did you use to transfer the technology?

Sugiyama: Sometimes R&D engineers and division engineers got together and had a dinner in order to communicate with each other. In addition, we had presentation meetings to show the results of R&D to division engineers twice a year.

Matsumoto: Dr. Sugiyama had some difficulty and he also mentioned that the managers in the R&D division were also faced with the same difficulty.

Sugiyama: Yes, with many companies including NEC and Sony—we had a summer meeting with persons of the same position to discuss how to do technology transfer which was sponsored by Japan's Techno-Economics Society in 1970. But we could not find a good answer.

Aspray: I see that in 1975 you received the Purple Ribbon Medal from the Japanese government. Was that for all of your earlier work on inventions, or was it for your management work as well?

Sugiyama: I don't know what is the reason I received this medal. But I believe this medal was usually awarded to excellent inventors. As you mentioned, I had already applied for more than one hundred patents at that time.

Aspray: One hundred by that time.

Sugiyama: The total is 110, and around 1975 it was 80.

Aspray: Please tell me what happened in your career after that.

Sugiyama: I had stayed as the manager of R&D for three years and then I moved up to function leader in 1978.

Aspray: I see—for all of the corporate functions.

Sugiyama: Yes.

Aspray: So you had administration and marketing and R&D, and whatever else?

Sugiyama: It was 1978.

Aspray: Your title then was Executive Vice President?

Sugiyama: Yes. I wanted to expand Yokogawa business to be not only in control but also in other new fields. I tried to expand into office automation, OA.

Aspray: What kinds of products in office automation?

Sugiyama: Japanese word processor.

Aspray: Yes. It's a very difficult task to build one, isn't it?

Sugiyama: It's a very big job, but I believed we could solve that. Many companies, including Sony and Canon, joined in this office automation field. They were very strong and we stopped selling this kind of automation.

Aspray: They were much larger companies and had more resources to put into this, I take it.

Sugiyama: Yes. But the basic idea was from Yokogawa.

Aspray: I see. Did you try some other areas also?

Sugiyama: Yes. One was the medical business, but I had very severe experience in the medical field. For example, we developed some new type of cardiograph in 1965, but we had only hardware introduced into the medical field. Doctors did not accept our hardware. "What's this Yokogawa company," they would ask. We found that we needed some marketing first and then we could introduce hardware. This is very important. Around 1976 Yokogowa Electric Corporation and the General Electric Company signed a contract to sell medical instruments.

Aspray: You received a contract in 1976 to do some work on medical systems.

Sugiyama: Yes.

Aspray: What was the nature of this contract? What were you asked to do?

Sugiyama: Sales contract.

Aspray: It was sales. So they were supplying the technology and you were the introduction to the Japanese market.

Sugiyama: Right. Around 1975, GE wanted to have a strong sales network in Japan. Almost thirty-five companies proposed to sell the new CT system in Japan.

Matsumoto: In 1975 they opened this operation. They wanted to sell the product inside of Japan, and thirty-four companies applied to be their agent. Yokogawa was the last one to apply.

Sugiyama: When General Electric executives visited Yokogawa they were impressed with Yokogawa Electric very much because Yokogawa had a very strong engineering power and had servicing capability.

Aspray: I see. So because of the strength of the engineering group you could provide the technical service that was needed.

Sugiyama: Yes. And from 1976, Yokogawa started to sell GE's X-ray CT in Japan.

Aspray: What was the range of products?

Sugiyama: Only the CT. Yokogawa Electric Works caught up the network over all hospitals. Many doctors suddenly knew what Yokogawa does. And then new joint venture began in 1982.

Aspray: Why was it decided that the relationship should change and there should now be a joint venture rather than simply a contract?

Sugiyama: Good question. At the starting point, GE's CT system was selling very well, but Japanese companies such as Toshiba and Hitachi began to make not so expensive, small X-ray CTs, which they introduced in Japan. And our market share of the industry was beginning to drop rapidly. Yokogawa made up our mind to build some new CT for the Japanese market. At that time Mr. Sugita was in charge of this business. GE found that Yokogawa was now making a new CT system. GE said, "It's not good." If so, we should have some new joint venture that will sell both GE's one and Yokogawa's one.

Aspray: Did GE find it was a very good system?

Sugiyama: Yes. Syozo Yokogawa, President of Yokogawa, decided to have a joint venture.

Aspray: What was the ownership in the venture? Who owned what portion?

Sugiyama: GE had 51 percent and Yokogawa 49 percent.

Aspray: You were appointed to be the president when it was founded?

Sugiyama: Yes.

Aspray: What were your main challenges getting this new business going?

Sugiyama: Frankly, I did not like this kind of arrangement.

Aspray: I see.

Sugiyama: But Shozo Yokogawa said, "You may have a good experience as a president of a joint venture. You may learn many things from GE." After a week I said, "Okay, I will do it as the president." The engineers, total number of employees was 333. Yokogawa Medical Systems started in 1982.

Aspray: Were the engineers brought over from Yokogawa Electric?

Sugiyama: Yes. Sixty engineers joined the new company.

Aspray: I see. What did you get from General Electric? What did they provide? Did they provide any technology, or did they provide any financial help?

Sugiyama: Technology came from GE, especially since GE has a very nice software system. Premium systems were from GE. Conventional type ones and medium ones were from Yokogawa.

Aspray: So that the top of the line was provided by GE and the rest of the products and product line was produced by Yokogawa?

Sugiyama: Yes.

Aspray: Did they share lots of parts? Was there that kind of cooperation?

Sugiyama: Components were separate. Then our products were exported to the United States using GE's network. It was a kind of OEM [original equipment manufacturer].

Aspray: So they slapped their name onto your product, and used their marketing abilities and connections to sell it? But that was a good test of how it stood competition in those places. Right?

Sugiyama: Yes. It was the same system that we used with MRI. The premium line was the GE line. We had a very strong team compared with Toshiba or Hitachi or Siemens.

Aspray: Just to make sure that what you wrote on the board gets onto the tape, so for MRI the same thing happened. The top of the line product was produced by GE and the others were produced by YMS. That made for a very strong cooperative venture product line, better than Siemens and some of the other competitors.

Sugiyama: Yes.

Aspray: Who were the other competitors besides Siemens?

Sugiyama: Philips and Picker, an American company. Philips is in Europe. But Siemens is very strong. GE is number one and number two is Siemens. Philips and Picker have smaller shares.

Aspray: I see. Did the product line grow over time? I know now that you have a large number of products on the market. What was your strategy for growing the business?

Sugiyama: The growing speed was faster than I expected. My plan was always delayed because the Japanese market was almost saturated with CT but is now expanding especially in the Asian division, especially since China and India want to have this kind of CT.

Aspray: Lots of people want it. As soon as they can afford the advanced medical equipment, there's a great need for it.

Sugiyama: Yes. We hope to expand our factory, especially low-end CT and MRI systems.

Aspray: But you also introduced new product lines, ultrasound for example, and were there some other new product lines as well?

Sugiyama: We have ultrasound equipment, but the total sales volume is small because one unit price is not so high. Maybe one-tenth of CT's price.

Aspray: I can see how the products fall into the same kind of market. They're used for similar kinds of things by the users. Are the technologies that they're based on similar enough so that you could easily transfer your knowledge from building one kind of system to building another kind of them, from going to CT to MRI to ultrasound, for example?

Sugiyama: Basically, the MRI and CT and ultrasound are different technologies. But concerning MRI and CT, the console desks are using almost the same technology.

Aspray: So you did have some economies of scale in the engineering in a way?

Sugiyama: Yes, that's right.

Aspray: How did you manage the rapid growth? What kinds of techniques did you use to enable yourself to grow so fast? You had to increase your number of employees and you changed buildings frequently. How could you make this rational and profitable as you grew so fast? What techniques did you use?

Sugiyama: When the number of engineers is growing and they have been so busy, the employee's mind is very excited and there is no problem. On the other hand, when in a recession, the engineer is not so busy, is not excited, is discouraged, and is doing other things. Maybe the engineer cannot concentrate on how to make the target date.

Aspray: As we were walking through the factory, you mentioned that you used a just-in-time system that was introduced by Toyota.

Sugiyama: Oh, yes.

Aspray: Was this something that was widely known or did you actually have to get assistance from Toyota or somebody else to set up your system for you? Did you go outside and get experience or was it just that it was a known system?

Sugiyama: I want to explain. Around, I think, fifteen years ago, at Yokogawa Electric Works, Syozo Yokogawa came to know a person, Mr. Kinoshita. He established a new concept of a manufacturing system and some instructor joined this group. Its name was the NPS group, the New Product and System group. Many of the instructors came from Toyota. He brought some new production techniques into this group. His group taught us through lectures. We have learned from this teacher in Yokogawa Electric Work. YMS is using almost the same technique, this is basically Toyota's Kanbom System, the just-in-time philosophy.

Aspray: Were you able to implement it smoothly? Were there major problems in introducing it into your business?

Sugiyama: It was very difficult. Yokogawa Electric Corporation is an old company, especially the manufacturing employees are very against this new technology.

Aspray: But it should have been easier to implement in a new company even if you had engineers that came over.

Sugiyama: That is right. At this time Mr. Katagiri, who is a very strong leader, imposed his idea based on NPS with his efforts. After ten years, Yokogawa inventories go down. Maybe one-third or one-fourth of them.

Aspray: Oh, a very great savings of money that way. So you've been very satisfied with the use of this system within YMS.

Sugiyama: Yes. Now I was satisfied with the system. However, recently I introduced a new system, a TQC system—total quality control system. My friend, Mr. Uchimaru, who was a president of a NEC subsidiary, was retired, and he wanted to coach our company. An NPS system is only for manufacturing, but a TQC is a total quality control system.

Matsumoto: Quality control is based upon the new technology for management. They propose activity in the company from the sales side to production to engineering and so on. So the system is not only limited to TQC, but also includes all the other activities—we have some directive measures and control measures. This time the company decided, that is, the employer decided, to employ total quality control.

Aspray: I see. When did you begin to implement this?

Sugiyama: Two years ago YMS started policy management based on TQC. At first the president shows *policy management* (Hoshin Kanri), then, it is broken down to the manager's level.

Aspray: What other important technology management lessons can we learn from YMS? What sorts of challenges did you have? What sorts of successes did you have? What were some of the big management issues that had to be faced in YMS and how were they resolved?

Sugiyama: I didn't have a big management problem.

Aspray: I see that in 1988 you were appointed as chairman.

Sugiyama: Yes, that's right.

Aspray: What was the significance of that change? How did your duties change? How did it come about?

Sugiyama: GE wanted YMS to grow more internationally. If so, the president should be from the GE side. GE decided this. GE already established a three-pole system. One pole is from Milwaukee in the United States. GE CGR is in Europe and

Sugiyama: in Russia. Yokogawa Medical System is the center of Asia. That is a three-pole system.

Aspray: I see. So you divide the world into three parts and each of these divisions is responsible for that whole area.

Sugiyama: Mr. John Trani was appointed as a division manager of Milwaukee. This is the number one reason the three-pole system is established. The chairman should be American or European. The second reason is stock share changed. Originally 51 percent is GE, but Jack Welch, the chairman of GE, required Syozo Yokogawa to change the share to 75 percent GE and 25 percent Yokogawa Electric Works. That means the interest of GE is getting strong.

Aspray: Yes. What were your specific duties when you became chairman? How was labor divided? What responsibilities did you have?

Sugiyama: In Japan, the chairman's duties, in general, are always lighter than the president. Mainly, I went outside and I met some hospital presidents or some major professors, and I went to the United States and other countries. The new president, Chuck Pieper, concentrated on company needs inside the company.

Aspray: I see. So, the president was involved more with the everyday operations of the business?

Sugiyama: Yes. Daily operations.

Aspray: Did the chairman's duties involve long-term planning for new business areas?

Sugiyama: Yes, and the chairman is director of the board meeting. The manager meeting is the president's responsibility. I don't attend the manager meetings.

Aspray: Does the president sit on the board though?

Sugiyama: Yes.

Aspray: You held that position for three years?

Sugiyama: Yes. Three years.

Aspray: Then you were appointed as senior technical advisor. Is that right?

Sugiyama: I was assigned as an advisor for YMS and an executive technological advisor for Yokogawa Electric Works. I have two hats now.

Aspray: I see. Since 1991 what have your main responsibilities been? What are you looked to do as advisor?

Sugiyama: I come to YMS two days a week and mainly a manager comes up to me for everything. For example, we are now starting some company merger. I was consulted about the strategy. Then the president goes to this company to discuss it. This is one kind of task.

Aspray: You spend your other three days over at Yokogawa Electric?

Sugiyama: Yes.

Aspray: What are your duties over there? Same sort of things?

Sugiyama: Something different. In Yokogawa Electric Works, I mainly concentrate on technological items.

Aspray: Let's discuss the question of capital. In YMS you have this rapidly growing business. Where did you raise all the money to hire these new employees and build these new plants and expand so quickly?

Matsumoto: He's not sure that this is accurate but he imagines all the earned money was gained from the profits.

Aspray: So you're able to build from your revenue?

Matsumoto: Yes.

Sugiyama: Also borrow money from bank.

Aspray: To get a line of credit from a bank, did you have to put some bank officers on your board of directors?

Sugiyama: No.

Matsumoto: You see, it has to do with the profitability of YMS because, as you know, a bank does not want to loan to a company that does not produce profits.

Aspray: But sometimes, especially with small companies, it's frequently common that bank officers worry about the management of the company because it's growing so rapidly. So they'll want to make sure that their investment is protected and they want a say on the board of directors. That happened, for example, recently in the Ascii Corporation. So one of the commercial banks in Japan now has, I think, two seats on the board of directors. But that was not the case here.

I have a question about technical background. Either in YEC or YMS, when you come to a major decision, are they typically business decisions, are they technical decisions, or are they some combination of business and technical decisions? If they involve technology, what is the process used? Do you get advice from your senior R&D people about the decision? Just how does it take place?

Sugiyama: When I was the president of YMS, the main decisions were made by the R&D manager. We only talk to him. Other business decisions were made by myself.

Aspray: I guess it's more a business decision than a technical decision to enter a new field, for example, to start your ultrasound business.

Sugiyama: On big decision making, we need some discussion with GE. For example, we have many kinds of MRI systems. There is a very strong one, 1.5 Tesla, made by GE. YMS should ask what kind of strength their MRI system should have. It's a very important strategy. We had to meet many times with the GE engineers and management. We flew to Seattle or Hawaii. It's very important. It takes around one year to make that decision. Once decided is not so difficult to the Yokogawa Medical Systems.

Aspray: My next question concerns the kind of qualifications needed to become part of senior management. Suppose that a young engineer comes to work for the company and seems to be quite promising. Do you take special steps to move him or her around in the company to get the right background so that he or she is well prepared for a senior position? Do you purposely move them around?

Sugiyama: At this point, GE has a very good system. We are now learning from this system at GE. GE's career-moving system. According to this system, for the first five years, the engineer stays at, for example, ultrasound technology. If he is a very excellent engineer, he could move to the MRI system, and after ten years he'll be a manager of this section.

Aspray: In the United States, there's so-called fast-tracking. That's where you move people through the system much more rapidly if they show great promise. It seems like it would be much more difficult to do within a Japanese company given the concern over the group of people that come in at a certain time and so on. Is that true, that it's more difficult to do fast-tracking in a Japanese company?

Sugiyama: Yes. In Japan, it's very difficult because this kind of rapid jumping is not good for teamwork. But recently we are now introducing the American way. The American system is much better now.

Aspray: I see. Are there ever examples of forty-five year-old vice presidents? Could somebody move up the ladder that quickly?

Sugiyama: Around ten years ago, it was very difficult in Japan for this to happen, but recently it has begun changing. For example, our managing director is forty-eight years old. But this causes no problem.

Aspray: Let me turn to question five [in your written answers] about quality, maintainability, and reliability in products. I can see that in a medical system reliability is quite valuable, quite important.

Sugiyama: Yes.

Aspray: What kinds of methods did you use to ensure the quality and reliability of your products? Did you use special techniques, whether they are management techniques or technical techniques, did you put in extra relations with the customer, or just what kind of things did you do?

Sugiyama: One thing is recently introduced, some customer satisfaction movement. It's very important. We meet many customers, not only the doctor—also some nurses and some technicians. They ask what is our product quality. Sometimes they complain that "This is not good." If so, some engineer visits this hospital and discusses with this technician to understand what happened in this hospital and improves some service. It's very important to contact hospitals directly and to hear their complaints.

Aspray: What about assembly? How do you assure quality of assembly?

Matsumoto: I'll explain the basic idea. It is basically the idea of just-in-time. The just-in-time basic idea is to make some required component only when it is required. In the old-time manufacturing industry, the employee usually makes ten or twenty quantities at a time. But under the just-in-time system, he makes only one part when required. From the standpoint with the production engineering, this is the old-time manufacturing start. First ten or twenty units are sent from the outside for cutting. Then they are sent to the storage man. Then process two people receive the three materials at that time to make an assembly. You can see this is old type of manufacture.

Aspray: That's right.

Matsumoto: But the new type of manufacturing based on just-in-time is a process where people receive one quantity from the store.

And then, after the cutting has been made, they are sent to the center and processed.

Aspray: So it doesn't go to storage, and so on.

Matsumoto: Thus the elapsed time is shortened from here to here and from here to here. This is from the standpoint of the production engineering. How about from the standpoint of quality? This is a lot of production over the conventional way. For instance, one receives the three materials. Then process-one people make some assembly or some cutting and then send to storage between process one and process two. But when the process-two people find some difficulty or some malfunction, they have to know, "What was the reason?" But already the time has passed from process one to process two.

    Because of the time lag the process-one people sometimes forget. They need to find what's the real reason. But, this is the case of one by one production. From the process-one people, the process-two people receive the cut material. When they find some difficulty or some faults they can make some necessary feedback at the time they find it. So they can take some necessary preventive action. One by one production makes the best quality. This is the same idea on which this process was carried out. So every assembly will be checked at the time by next process people.

Aspray: I also noticed this morning when we were touring there was a place for customers to come in to check over their own products before they went out the door. Is that so that things can be fixed while they are there rather than be sent to the site and have to be sent back and such? That seems to be another step toward quality.

Sugiyama: Yes.

Aspray: You mention in your answer to question number 10, the Tollgate system. Could you explain what that is to me?

Sugiyama: The Tollgate system is for the development process. There are many kinds of checkpoints. The Tollgate system is like paying money as you are going through the road. At each section we have some meeting and explain the development process and all managers say, "Okay, go to the next step."

Aspray: So this is part of the research planning process.

Sugiyama: That's right. It's the GE Tollgate system.

Aspray: I assume that that's a way of preventing a research group from going off in one direction for a long period of time, without staying on track for the company's needs.

Sugiyama: We need some sort of a checklist.

Aspray: Are there some sorts of group decisions made in terms of reviewing R&D programs and projects, or is it a particular manager that makes decisions? Who makes the reviews in the R&D system?

Sugiyama: It's a top manager of R&D who decides "go" or "no go." That's a meeting where some marketing people join the meeting. Because marketing people's opinions are very important.

Aspray: Right. So that you're doing research that's directed toward products that are needed.

Out on the floor there was a discussion with one of the people who showed us around, and she said there weren't distinctions made in his area between technicians and engineers. What's the rationale for that?

Sugiyama: Mr. Ibuka, former president of Sony, said the Japanese engineer has two hats. Sometimes he's a technician and sometimes he's an engineer. He said the Japanese engineer has two mindsets.

Aspray: Does that have a particular advantage in the engineering design process?

Sugiyama: I think, sometimes the engineer is only sitting at a desk and doesn't want to do any practical job. But the technician does some soldering, wiring. The Japanese engineer needs both sides.

Aspray: So you learn something from the soldering that you can use in your design work and vice versa.

Sugiyama: That's right.

Aspray: With regard to question number 13, on alliances with government, business, or academia, could you tell me more about how the relation with Keio University works. What is the relationship between the two of you, more specifically?

Sugiyama: I know many medical doctors at Keio University. Also, Shozo Yokogawa graduated from Keio University. And many of Yokogawa's directors graduate from Keio University. Yokogowa Electric Works has a close relation with Keio University. For YMS we need some testing site for new products. Keio University is very important for us and we want to keep this close relation. Keio University is now using our imaging system. Not those of Toshiba or Hitachi.

Aspray: Why would Keio University want to have you come there? What advantage do they get from the relationship?

Sugiyama: Keio gains some new technology from GE and information from GE. GE learns some medical information from American medical doctors, which is a very important information route.

Aspray: They also get to use the equipment at no charge in their research and in their diagnostics.

Sugiyama: Yes. Somewhat at a low price.

Aspray: So they like this as well. I suppose it also means that they get the newest technology?

Sugiyama: Yes.

Aspray: Do you actually have some of your engineers or research teams spend time at Keio University on site for periods of time?

Sugiyama: Yes. Always one or two engineers or servicemen stay in Keio hospital.

Aspray: On the next question, on qualifications for professional employees, you say that it's difficult to recruit highly qualified people. Is that because there's a shortage of electrical engineers today, or some other reason? Do you need to have medical people working for your company or is it sufficient to have only engineers? Do you need both?

Sugiyama: Only engineers. No medical doctors here.

Aspray: Do you have trouble hiring good engineers?

Sugiyama: Recently it has not been difficult because the company's name is getting famous in Japan.

Aspray: I was very impressed as I walked through the building to see how much space and how much time is devoted to continuing education. Why is it that you should devote so many resources to that? What makes it so important to spend a lot of money on continuing education?

Sugiyama: Our target is that 5 percent of total working hour is spent on continuing education.

Aspray: That's more than lots of other companies spend. Some companies do very little training of their engineers. You think it has great value to your employees?

Sugiyama: Yes.

Aspray: In planning a new product, what is the expected life cycle time—development time?

Sugiyama: It depends on product research. For example, CT is very short—only two or three years research time. In case of MRIs, it's maybe five years.

Aspray: So you have to introduce new products in those lines that frequently. Typically, within the company, are you working on more than one generation of products? Are you not only working on the next product but the product after that at the same time in some sort of way?

Sugiyama: Only the next generation's product.

Aspray: For the future of YMS, do you think that it's more likely that the company is going to grow by expanding into international markets with the same product lines, or produce new versions of the same product lines, or instead introduce whole new additional product lines? Which is your strategy for growth?

Sugiyama: We already established three poles: the CT, the MRI, and the ultrasound. Another one, the X-ray system, is still small. For brand new fields, we are now exploring a PET system (Positron Emission Tomography). It shows some kind of brain activity. But it is a very expensive system. We are expecting that new technology. Another one is to detect brain Magnetic Encephalogram (MEG), a kind of wave from the brain. It detects a magnetic field that we call the MEG. It is now in the research stage. We are conducting some tests at ETL (Electro Technical Laboratory).

Aspray: Could you summarize for me what you think are the great advantages, but also the great disadvantages of being in a joint venture?

Sugiyama: Good question. First of all, the advantage is that we receive new technology from GE. And our mindset is more international, maybe better than Yokogawa Electric Works. It is natural for us to have an international mind. Concerning export, is very natural to use GE system. As concerns the OEM export to United States, sales and service departments can use the GE system. This is a great advantage for my company.

Disadvantage. The market share change gives less influence to Yokogawa: 75 percent of profit goes to GE; only 25 percent goes to Yokogawa.

Aspray: I see. So you really have to make a lot of money for Yokogawa Electric to make out well. Are there disadvantages over management styles or over those kinds of issues?

Sugiyama: Another point is it takes so much time to decide some kinds of issues. For example, we want to expand a factory some other place. We need permission from the GE side and the Yokogawa side. And it takes a lot of time. Sometimes policy is different.

Aspray: Are there very strong differences in the corporate cultures of the two companies that make it difficult for you to work between them?

Sugiyama: At the first stage, I felt it strongly, this kind of corporate culture. Now I don't feel so.

Takashi Yamanaka

Place: Tokyo, Japan

Date: February 18, 1993

[*Note:* Mr. Yamanaka speaks in Japanese, and then some of his words are interpreted by Mr. Eiju Matsumoto, curator of the Yokogawa Museum Office.]

Aspray: I would like to begin by discussing why it's important to have an engineering background in order to manage a company. I'm not, however, a business-type person. I don't come out of a management school or a business school. I'm professionally a historian. So I believe that one can learn lessons from past episodes in a company's history. And I hope that in the interview today that we can talk about general issues that are important in the management of a company, but tell them in the context of past episodes in the company, particular examples from which you learned in some aspect or another.

Yamanaka: First let me talk about the inauguration of Yokogawa. Yokogawa was established by a person named Dr. Tamisuke Yokogawa, who had been a famous architect in Japan. That was about seventy-eight years ago. He had built many big and famous buildings here at that time, and he was the chairman of the Architecture Society. He had succeeded not only as a scholar, but also as a businessman. He was very much interested in the fundamentals of those technical things, including the electrical power supply, and the construction of the buildings in which iron frames and steel bars were used in concrete. He couldn't find any kind of instruments in Japan, so he decided it is important to learn these techniques from European countries.

Dr. Yokogawa made several companies. One of them was Yokogawa Electric Company, and it started from the domestic production of instruments. At that time almost all these instruments were imported, and we couldn't find any kind of things for production.

Aspray: Were there other instrument companies that were competitors at the time?

Yamanaka: At that time there weren't any competitors in Japan. Almost all of these instruments were imported from General Electric, Siemens, or Westinghouse.

Aspray: I see.

Yamanaka: Dr. Yokogawa employed four people and started Yokogawa Electric Meter Research Laboratory.

I must mention the importance of the philosophy of the establishment that Dr. Yokogawa made. It included three items: One is quality first. The second is the pioneering spirit. Third is the contribution to society.

Nowadays, as you know, especially after World War II, quality management and quality control are important. That is something every person can acknowledge. However, at that time, especially in the small enterprises such as the Yokogawa Electric Meter Research Laboratory, putting quality first was a very remarkable and epoch-making thing.

Aspray: Indeed.

Yamanaka: Dr. Yokogawa insisted not to plan only on profits, but for supply and high-quality products for society.

His second philosophy was to have a pioneer spirit. He intended to produce creative instruments, which were not merely a modification of competitors' instruments, nor a modification of the imported instruments.

The third item was his contribution to society. At that time Japan was in a growth period for capitalism. The main employers at that time insisted that their employees simply be good employees. But there were few employers who educated employees to have a sense of contribution to society through supplying quality instruments.

All three philosophies remain in the vision of Yokogawa, among all employees. For example, the contribution to society means that all the employees have to be good citizens before becoming good employees. They serve the philosophy by contributing to the regional areas.

Through a recent corporate identity project, Yokogawa established two company spirits and the company gained. First, Yokogawa has to develop products concerning the fields of measurement, control, and information. Second, Yokogawa has to be a good citizen and play the role of pioneer spirit. Inaugural company philosophies developed into today's company spirit. From these company spirits, we established a company vision. We have long-term economic prospects. And after the long-term prospects, we establish medium-term prospects. In one or two years our activities show how we are making progress. On the basis of these three prospects—long-, medium-, and short-term—we usually make some estimation of the next fiscal year and make necessary adjustments. This is the internal organ called "Executive Form." (Yamanaka explains using a company organ.) This is restricted or limited to all the Yokogawa internal, top management.

Aspray: I see.

Yamanaka: The corporate philosophy and the corporate vision, together with the structured business plan, make up Yokogawa's management core. This is supported by employees' expectations, top management's philosophies, and the inherited characteristics and history. I already explained the foundation and history of Yokogawa.

We have to keep this fundamental spirit of the foundation and it changes in the nineties. The nineties are very, very difficult years, and the many changes are so rapid. The company wonders if we should follow the big global change. That is our Vector 21. This is a long-term management concept which provides direction to our year 2000 market portfolio. It also outlines the aims and the duty of each segment and division. I explained these directions not only to the domestic company, but also to the global-based affiliates and factories worldwide. Yokogawa has about seven thousand employees in Japan and another three thousand employees outside of Japan. So about ten thousand employees of the enterprise are global-based in Yokogawa. We have mapped our direction to the year 2000. The management philosophy of today is based on the Yokogawa Vector vision and also the long-term management vision. I'd like to talk about how to manage an overseas company and overseas employees.

Aspray: Very good.

Yamanaka: Many people think in managing overseas subsidiaries they should send the dividends from the interest of the subsidiaries to the parent company. But Yokogawa focuses on other priorities in their subsidiaries all over the world. First, we advise them to seek successful business in their countries. Do not produce only the designs by Yokogawa but also the individual designs by the people in the area, and make those products in their own countries. Then enlarge the employment. From the interest they earn in their countries, they invest in their own countries. And last, send your interest to Yokogawa.

As I mentioned before, all of our employees have to be good citizens in their respective countries before they are employees of the company. Our philosophy is based, as I mentioned before, on the inauguration spirit of the company. For instance, in the overseas subsidiaries, the necessary thing is to be acknowledged in their domestic countries and so contribute to their countries first. Their company also has to be a good citizen in the home countries.

Next I have to mention the relation that we have with the Hewlett Packard Company and also with General Electric. We established a joint venture with the Hewlett Packard Company called the Yokogawa-Hewlett Packard Company. The Yokogawa-Hewlett Packard Company contributed not only to Yokogawa but also to Hewlett Packard. They contributed to the interests of the United States.

Before we established a joint venture with the Hewlett Packard Company, Hewlett Packard was the biggest manufacturer of electronic measuring instruments. However, at that time, their sales amount in Japan was only two hundred million yen. However, after we established the joint venture with them, we transferred many employees to the joint venture. Now Yokogawa-Hewlett Packard has sales that amount to more than sixteen billion yen. The company has the top priority for engineering and the top share of electrical measuring instruments.

Aspray: When the joint venture was first established, what did each company hope to gain from the joint venture? And in practice, over time, did they actually gain those things they hoped to gain?

Yamanaka: Especially at that time in Japan, Yokogawa wanted to employ high-frequency measuring techniques. Yokogawa wanted those kinds of technology and wanted to get some market share in the field. On the other hand, Hewlett Packard had a very low market share and wanted to increase its sales in Japan. They expected Yokogawa to assist in the field of measuring instruments and to provide strong sales and service facilities. So the two companies could teach each other; they had complementary strengths to offer one another. After the establishment of the company, which happened more than thirty years ago, they have attained the position of top manufacturer and supplier in Japan in their market area. In other words, the joint venture contributed to both of the parent companies, and the joint YHP is the typical example of success.

Aspray: Over time, the ownership of the two companies in this venture changed. Is there some significance in that?

Yamanaka: Originally, Yokogawa held 51 percent of the shares, and Hewlett Packard held 49 percent. However, about several years ago, the ratio changed to 75 and 25 percent (75 is Hewlett Packard Company). There were many reasons. One was from the standpoint of Hewlett Packard Company. They did not have any other joint ventures except Yokogawa-Hewlett Packard in Japan. All the other factories in the world belonged to Hewlett Packard. So the Hewlett Packard Company wanted Yokogawa-Hewlett Packard to become the center of its Asian activities. And to become the company to contribute more to the rest of the world. With Hewlett Packard as the major partner, they can inject more resources than before.

Aspray: How would you describe relations between Japanese and American companies in general?

Yamanaka: As you know, between Japan and the United States, there is some dispute. A particular example: Japan has been introducing technology from the United States, and they improve it on the basis of production techniques and quality. Japan intends to make many instruments and many products for the United States, and does not want to import any products from the United States. As I mentioned, the Yokogawa-Hewlett Packard Company is a good example because it contributed not only to Yokogawa, but also to Hewlett Packard. Finally, they can sell more than ten billion dollars of instruments in Japan. From the global standpoint, that is an example of good success in a joint venture.

On the other hand, Hewlett Packard Company has provided very good management concerning a Japanese company. They respected the Japanese management philosophy and management ways. For example, the president and the vice president working in the Yokogawa-Hewlett Packard Company have always been Japanese. The director and the board has representation from the Hewlett Packard Company. But there are only two persons from Hewlett Packard who attend the Yokogawa-Hewlett Packard stockholders meetings twice a year. They are not exclusively working for the YHP. So that even though the capital ratio is 75 to 25 percent, Hewlett Packard decided all the management will be done the Japanese way.

Another example is the relation with General Electric Company, especially in the medical division. Up until now, General Electric had the intention to sell these medical products in Japan. However, it was very hard to become successful. They concentrated their efforts on imaging equipment and established a joint venture with Yokogawa.

In regard to the equipment, I refer to computer tomography and magnetic resonance imaging equipment. Within five years of the joint venture, the joint venture company had one of the top shares of the computer tomography field. There are other companies—for example, Toshiba and Hitachi—that are into those kinds of equipment businesses. Even in the case of General Electric, they have different business. However, they didn't succeed in a relatively short period, nor did they gain a top share of the market. The famous Jack Welch, chairman of the General Electric Company, used to say that there is a lot of friction between Japan and the United States. One of the examples is trade friction. However, we don't have any friction between General Electric Company and Yokogawa because we are tied with mutual reliability and mutual interest.

Aspray: From what I've read about this joint venture, Yokogawa's role has been far more than simply marketing and manufacturing locally. They were thoroughly involved in product development and product enhancement, and there was a technology transfer back to the United States that was very important in the American market. Could you talk to that?

Yamanaka: This is my opinion: America is traditionally good in the field of innovative new technologies. On the other hand, the Japanese are used to being good in the field of production engineering and quality management. So at Yokogawa Medical Systems, we imported American technology at first to be our computer tomography. However, as a result of our joint endeavor, focusing particularly on the field of production engineering, we were able to develop low-cost equipment and also provide quality management. We built a Japanese version of the computer tomography system. Now we and General Electric can export to the rest of the world. This describes the relation today.

One of the relations between Yokogawa and overseas countries is based partly on the philosophy of Shozo Yokogawa. He is one of the children of the founder and now he is the chairman of Yokogawa. His idea was based on the global view and three items from inauguration spirits I described earlier. Yokogawa has to be and think not only of Yokogawa, but also of the counterparts all over the world. So the good relations with Hewlett Packard and with General Electric Company are typical examples based on his idea and others of the company's inaugural spirit.

Another example is the relation between Yokogawa and China. We have two joint ventures with the Chinese government, and in a licensing agreement for technology we have seven licenses. One joint venture is focused on the process control industry, which manufactures from analog type control systems to distributed control system. The company, located in Xian, has the top share in the field of industrial process instruments in China. Another company, located in Suzhou, focuses on the field of indicating meters. The company produces one million units a year. All of the units are exported to Japan and the United States. They also get the standard certificate from Japan.

There are few companies that produce technological products on the basis of the recent technologies exported from the

United States as well as Japan. Just to give one example: It is said that China doesn't have any high technologies. However, after we thought and transferred lots of the know-how to the joint venture company, they become able to make their own instruments. The company has contributed to the Chinese government and was honored with an award as one of the top export companies in China.

I have mentioned examples of companies in the United States and China. Some people are now worried about the boomerang effect. Do you know the boomerang effect?

Aspray: No.

Yamanaka: The boomerang effect is experienced when a company which has high technology exports or transfers the know-how to the not-advanced country, such as China. Eventually, the flow will go back "by the boomerang." This is what we called the boomerang effect. There are people who warn about the boomerang effect, and they do not want to export or transfer technology. However, Yokogawa doesn't think of those worries, and we always exchange the know-how, even in China. We believe those companies should be good and should contribute to their own countries.

Contribution to society is the idea of Dr. Tamisuke Yokogawa. Dr. Yokogawa, who founded Yokogawa, was famous as a collector of china—old Chinese china. His collection was equivalent to the collection of Mr. David in England, who collected more than a few thousand items. After he collected the china, he donated his collection to the National Museum of Japan. It is said the price was more than twenty billion yen. Now, if you visited there you could see his collection. His donation to the National Museum symbolizes his company policy. When he inaugurated Yokogawa Electric, he had an idea, he had a philosophy: He didn't want to have company shares. He believed the company belongs to the public. Now his china collection is also available for the public to enjoy.

Aspray: I hope to see the collection.

Yamanaka: I mentioned the history of Yokogawa and its philosophy. Yokogawa has a philosophy about the field of technology and about products. First, we don't want to modify an existing model of the other competitors for our other products. We don't want only to modify. We want to create.

Aspray: I see. You don't want to simply take and copy other people's products.

Yamanaka: That's right.

Aspray: You want to create your own.

Yamanaka: Yes. One of the examples in the field of process automation emerged when Yokogawa first announced the idea of the distributed control system, which is sold to the factory. At that time, the only competitor was Honeywell. After Yokogawa announced this system in the works, Honeywell announced a similar kind of instrument, but five months later.

Aspray: What year was this approximately?

Yamanaka: The announcement was made in 1973, and we started to sell the system in 1975. However, when you have a chance to visit Honeywell and ask the same thing they might say Honeywell was first and Yokogawa followed after five months. Yokogawa has the top share in Japan in the market as well as in South Asia. We intend to expand our share of the instruments in Europe as well as in the U.S.

There is an example. This is a copy of a magazine named *Control*. There is an article which describes the result of a questionnaire in the U.S. The question is "How do you feel about the company or how do you feel about the level of the companies which reacted to the process industry?"

Yokogawa has a joint venture with Johnson Control in the United States called J.Y.C. We gained much knowledge from the questionnaire.

There are only limited products listed. However, the top Voltex flowmeter is from J.Y.C., which gains 57 percent. Second is Foxboro Company with 26 percent. We got third in magnetic flow meters.

The other transmitter is the most popular in the field of liquid flow rate measurement and uses the principle of differential pressure detector. First is Rosemount with 55 percent, second is Honeywell, and third is J.Y.C. and Foxboro with only 5 percent. I remember twenty years ago Foxboro had a top share at 65 percent.

Concerning small-scale distributed control systems Johnson-Yokogawa has a top priority. However, in full-scale distributed control Yokogawa didn't deliver until now.

Aspray: I see. Are these market shares from the worldwide market or from American?

Yamanaka: We can get a market share report from Frost & Sullivan in world-wide categories. This is the U.S. We expect large marketing in the U.S. recorder business. Recorders are traditionally Yokogawa's strong techniques and strong market area.

Aspray: In the postwar period, there have been many technical changes—increase in digitalization, increase in the use of electronics, increase in the use of computers in your measuring fields. How has the company managed all of the change? What were your underlying philosophies? How did it cause problems for you, and how did you take advantage of it?

Yamanaka: In the field of process information, I remember that until the 1950s more than half of the instruments were pneumatic signal systems. However, all the techniques were changed to use vacuum tubes. Because of the lifetime of the vacuum tube devices the technology and components then changed to semiconductors. At the beginning germanium, and later silicon. That was about 1960. We employed integrated circuits including large-scale integrated circuits, LSI. Then in the 1970s we changed to digital techniques which were employed in DCS, Distributor Control System.

The most important thing for management is how quick the response is to the changes to the next imputative technology. For example, in the 1960s, more than half of the instruments, if not all of the instruments, were operated by pneumatic systems. At that time we had the license. We got the license from the Foxboro Company in Massachusetts. Yokogawa had the top share in the Japanese market. In addition seventy percent of the sales profit was gained from pneumatic systems. However, the most important issue for the decision making in management was to change those technologies from pneumatic to electronic. In the first stage, we expected to get red figures from electronic techniques. How management decides was a serious matter for us.

At that time Yokogawa delivered DCS in 1973. Almost all instruments were analog control instruments. Most of the profits were gained from those kinds of systems. But we have to change analog techniques to digital to develop DCS. We decided to put all the engineering forces in digital techniques. That was a very significant decision, however, and as a result we reached the status of Yokogawa now.

Aspray: Did that require either hiring a new set of engineers with different training or undertaking a series of continuing education programs for your older engineers?

Yamanaka: As you know, in Japanese society it is impossible to fire employees. We had many engineers who majored in mechanical engineering at the time of pneumatic systems. In other words, it is not possible to hire electrical engineers without using mechanical engineers. The same thing happened when we wanted to change analog techniques to digital techniques. We carried out re-education, continuing education of old engineers. In addition we hired new employees, particularly new graduate engineers.

Aspray: Did this also require an increase in the amount of money the company spent on research and development? I know that today your percentage is very high—over 9 percent.

Yamanaka: Before I became the president, investment in research was under 6.5 percent. Recently, after I was promoted to president, I wanted to increase it to 10 percent. Now I think in the previous year the ratio was 9.4 percent. That is a very good figure considering the business recession. The research and development expense has increased a lot, and at the same time we lost a fair amount of the profit. The necessary thing for me is to decide on the expenses for the future.

Aspray: What are the basic costs of developing new products? Is that one of the driving reasons to expand internationally, so that you can have larger markets to amortize your development costs?

Yamanaka: Are you asking if our increased R&D costs from developing instruments are recovered by exporting to the rest of the world?

Aspray: The degree to which your R&D costs increase, does that mean that those costs have to be passed on to the cost of your product? I would assume you have to have a larger market to spread them over. Is that one of the driving reasons for opening new markets in different countries?

Yamanaka: It is also said of General Electric products that, especially in the field of measurements, the change in technological innovations are rapid. On the other hand the quantity to be sold is relatively small. So the investment costs are high. When we want to develop some new product, it would often cost so much that it would be impossible to sell it in Japan only. So we have to see the conditions in the rest of the world. As a result, we do not have any kind of product that is exclusively developed for the Japanese market.

Aspray: Earlier you mentioned the example of establishing Yokogawa in Singapore. You talked about the hopes that they would not only develop marketing and manufacturing capability but also eventually be able to establish their company in that marketplace. This is a kind of decentralized strategy. Are there some aspects of the company that you think should remain centralized? For example, should R&D or legal concerns remain centralized in the company?

Yamanaka: The field of fundamental research I think, should be regarded at Yokogawa as a common factor. However, individual instrument design, especially relating to regulations of a particular country, should be done in each of the subsidiaries. Sometimes those companies have to adapt or change on the basis of domestic request. There are many examples. For instance, the products we produce for the U.S. or for Germany have to have some differences in their dimensions. Another matter concerns safety. Since industrial instruments are used in factories they should meet not only human physical safety standards but also must be explosion-proof. There are so many kinds of explosion regulations. I have heard there is a difference between Canadian, U.S., and German regulations. Therefore, products should be redesigned in individual countries.

In the future I believe even part of R&D should be redistributed. One example would be environmental measuring instruments detecting chlorofluorocarbons, which destroy the ozone layer in the atmosphere. I have heard the regulations concerning chlorofluorocarbons are stricter in the U.S. than in other countries, including European countries and Japan. I believe those special instruments should be designed in the U.S. in the future.

Aspray: I noticed when I was taking a tour of your research facility this morning, that the company had a policy of designing some semiconductor components that were proprietary. Can you talk to me about the strategy or the reasons behind that decision?

Yamanaka: As I mentioned before, Yokogawa Electric is not a major player in the computer field of technology. We are major in the field of measurements and control as well as information. So it is necessary for us to have measuring techniques for physical phenomena and to develop sensor technology. I believe those technologies should be done on the basis of these new electrodevice technologies and micromachine technology. The classical design techniques are not enough. It is necessary for us to adopt the advanced techniques. That is the reason why Yokogawa has a fair amount of semiconductor facilities as well as a research and development program facility. We want to develop fundamental devices as well as sensing devices. We don't want to make D-Ram, which are used for computers.

I want to mention examples. One is pressure sensors. We don't want to buy sensors from other companies, other suppliers. So we use a silicon diaphragm as a sensor and deposit the necessary process on the basis of silicon. Another example is high-speed A-D converters. When we purchase conventional A-D converters from outside suppliers, the result will be the same measuring instruments coming from another company. So we decided to develop them by our own technologies.

Aspray: Is there some effort to choose areas for basic research that will have application across many different businesses, that somehow you would want to get economies of scale for value? Most of the basic research that I was shown this morning in my tour had to do with fundamental technology that went into products themselves. Is there also research work on manufacturing technology for the company to improve all your economies?

Yamanaka: We have a different organization. Manufacturing technologies are done at the manufacturing engineering section and research group.

Aspray: It's thought in the United States—I don't know how true this is, and I'm sure it varies from company to company—that one of the great virtues of Japanese companies is that they have made very fundamental advances in manufacturing technology that have a great deal to do with profitability and reliability. Has that been the case at Yokogawa?

Yamanaka: Manufacturing technology is very important. We also have a mechanism called NYPS, New Yokogawa Production System. Most companies, as well as YMS, employ the production system which is based on Toyota's manufacturing method, just in time. This production system is one of the features of the Yokogawa group.

Aspray: How long have you used the system?

Yamanaka: About ten years.

Aspray: I have a question that is more general about product strategies. To a certain degree instruments are general purpose. They can be used in many different applications. But there is also a need in some cases for specialized or customized equipment. In which direction is the company's main interest? Is it in producing general-purpose instruments or in advancing that customized business? Or in no particular direction?

Yamanaka: Let me give some examples. As you say, with the general-purpose instruments, we can expect a fair amount of instruments to be sold. However, the specialized and the focused instruments that could be used in a particular instance, by a particular customer, can't be expected to have a large sales quantity. So in some cases we can sell maybe one or two units a month. That means in an ordinary case a decision will be made on the basis of whether it is profitable or not.

However, Yokogawa has been working in the field of measurement and control for a long time. I believe we have a great responsibility to supply standard instruments to be used in the areas of temperature, pressure, current, and electric power standards—even if the number we sell is small. Sometimes these kinds of instruments are employed as national standards in several countries in the world.

Aspray: We haven't really spoken about interactions that the company has with the government. In what way does the government play a role in the operation of the company, in determining the company's business? Whether it's standards, joint ventures, national products, military markets, or whatever?

Yamanaka: I would like to mention a few examples involving Yokogawa. Some company agreed to build developing projects. At that time we joined the national project. Recently the Japanese government started the Key Technology Promotion Center. The investment in this key technology was partly supplied from MITI and from private sectors for the rest. Yokogawa joined another kind of project, optical measurement technology called OMTEC, high-frequency technology called TERATEC, and Micro-Machine Project. The last one is the Super Conductive Material Project to be used in medical materials.

Aspray: This is the most fundamental question that I had for you: What importance is there in having an engineering background in being able to manage this company? There must be differences between running a technological business and running some other business that doesn't have to do with technology. How does having that experience in engineering affect the way you are able to manage a technological business?

Yamanaka: As I mentioned before, Yokogawa is a technology-oriented company, one that is based on the innovation philosophy made by Dr. Yokogawa. He intended to make a product, a new innovative product, which hadn't been seen in other markets. He made a word, "pioneer spirit." But on the other hand, from the managerial point, and from my experience, I believe the important characteristic for the manager is to have wider vision in the future. Another is to have the facility to decide the content of the information. I used to say, "To raise your antenna higher."

On the other hand, there are many engineers who don't have any sensitivity to the information. I think the person who doesn't major in the engineering field is sometimes more suitable than narrow-minded engineers. One of the features most engineers tend to hold is a lack of the wider views, as well as some particular thinking as demonstrated by the phrase NIH, not invented here. I don't want to use something because it's not invented here. That is a kind of narrow-mindedness.

Aspray: Let me give an example. One company I know of used to be a great technological company. But the management of that company decided that the technology didn't matter anymore. What really mattered was the short-term return on investment. So they would go wherever they could to get a good short-term return. That meant that they let their research operations drop off, that they diversified out from their strengths and technology because they were treating it like an ordinary business, not like a technological business. That seems to be one kind of extreme example.

There are many fundamental questions that have to be decided, issues that have to be decided, that involve technology. You have to decide whether you're going to start up a major research activity in a technical area. You have to decide what the future market is going to be for certain kinds of uses of technology and products. What is the role of the leader of a company in making these decisions? A person at this other company I just mentioned might say, "I don't have to care about the technology. I just look at my accounting statements and make a decision based on them." At the other extreme, for example, Dr. Kobayashi of NEC would say the most important thing I can do is have a technical vision of where the company should go. Where in that spectrum of alternatives do you think you are in your vision?

Yamanaka: Generally speaking, in Japanese society, shareholders evaluate the company over relatively long intervals of time. So even in a company that occasionally produces a deficit and has to reduce dividends, the top management are usually not forced to retire. As you know, in the United States in most of the companies who lose their profit and produce red figures the management are forced to retire. They have generally placed the priority on profit. I heard there are many instances of American companies who sell their own business to another company, and then go into a totally new business, changing from their traditional and conventional business. However, in Japanese society, those cases are quite rare. The difference between Mr. Kobayashi and the other company you mention is an example of the difference between the United States and Japan.

As I mentioned before, the Japanese circumstances tend to be oriented toward technological development. Even the technology for the government's military activities are operated by the research activities of private companies. Also the evaluation is made on the basis of your categories. I heard in the United States there are many venture business companies such as the companies in Silicon Valley. These have very challenging subjects. They are pursuing lots of new technologies as well as innovative seeds. On the other hand, those venture businesses are really supported by their different venture capitals. So I understand their opposite direction and opposite idea in the management of the big enterprises in the United States.

Aspray: Suppose your company is facing a major decision that involves some significant technical issue such as whether to fundamentally get involved in a particular new technology or new research program. I assume that the details of running the whole operation of the business don't allow you to keep completely up to date with the newest technologies. You have to rely on technical experts in your company. When there's a fundamental decision of that sort to be made, though, do you think it's important for you to take the time to really learn the ins and outs of the technical matters so that you can personally make the technical decision for the company? Or do you rely upon the director of the research and other technical officers?

Yamanaka: In ordinary cases, because we have lots of staff such as corporate R&D and marketing people we don't have any kind of problem making decisions. On the other hand when we face a big problem like entering a new business, a decision by top management becomes necessary. We have had examples of this in the past. Yokogawa has been working on meters and measuring instruments for a long time. We got a chance to enter the industrial process instruments area. Another one was at a time of technological change from analog control system to digital control system. In both cases we had to think seriously. Because most of the company's profits were gained from previous products, there was a possibility of losing a fair amount of profits.

Part III
United States

Chapter 9
Robert Galvin

About Motorola

Motorola began in 1921 as a storage battery company under the direction of Paul Galvin and his friend, Edward Stewart. The rise of electric power in the 1920s yielded a decreased demand for batteries, contributing to the company's failure only a few years after its founding. Responding to the changing technology, Paul Galvin and his brother Joe formed the Galvin Manufacturing Corporation (GMC) in 1928 to distribute a device called the battery eliminator. The eliminator, which enabled a radio to be plugged into an ordinary wall outlet, was distributed by Galvin's five-employee company to retail distributors such as Sears, Roebuck, and Company.

By 1930 Galvin Manufacturing Company had produced its first commercially successful car radio. It was this product line, under the brand name Motorola, that sustained the company during the Depression. GMC also sold police and home radios during the 1930s. By 1947 the company boasted five thousand employees. It purchased Detrola, a struggling automobile radio company, and subsequently supplied radios for all American Motors automobiles and half of those produced by Ford and Chrysler. Upon acquiring Detrola, Galvin Manufacturing Company officially changed its name to Motorola. A year later, Motorola formed a semiconductor development group and quickly became a major supplier of transistors to other companies.

The 1960s represented a turning point in Motorola history. During this decade, Motorola discontinued its production of consumer products, including radio and television receivers, to concentrate on activities in high technology. It expanded its operations outside the United States. In the early 1980s Motorola acquired Aizu-Toko K.K., renamed it Nippon Motorola Manufacturing Company, and began manufacturing integrated circuits in Japan. Motorola also purchased the computer terminal manufacturing company, Four-Phase Systems, in an unsuccessful effort to branch off into office automation and distributed data processing. In 1986 Motorola agreed with Toshiba to share its microprocessor designs in return for Toshiba's expertise in manufacturing dynamic random access memories (DRAMs).

Perhaps because of the company's relentless commitment to quality, Motorola has been extremely successful over the past decade. Their 68000 family of microchips, the cornerstone of personal computers and workstations by Apple, Hewlett Packard, Digital Equipment, and Sun Microsystems, was joined by their faster, more powerful 88000 microchip series in 1988. Sales generated by Motorola's six operating divisions (Communications, Semiconductor Products, General Systems, Information Systems, Automotive and Industrial Electronics, and Government Electronics) now exceed $8 billion. Motorola is currently the world's top supplier of cellular phones and was honored as the first winner of the U.S. Department of Commerce's Malcolm Baldrige National Quality Award.

Robert Galvin

Place: Schaumburg, Illinois

Date: April 14, 1993

Aspray: Perhaps the best way to start is with a very general question. Motorola has been extremely successful over the last few decades. Can you isolate several elements that you think differentiate your company from your competition? Or at least ones that are the keys to the success of the company.

Galvin: I'll start to develop that list, but I cannot do it in an orderly way with a hierarchical rating of these factors. What success we've had, I think, is a function of a culture of a very significant involvement on the part of We, W-E. It is not the function of one engineer or one or two people that have done a particular thing. It's a fairly broad spectrum. This has been helped by a spirit and a drive at the top of the company that we should leave relatively no stone unturned. Therefore there has been a high spirit of risk and commitment to what needs to be done on the part of the we's.

So we've had a very substantial number of people who in their field—whether it's semiconductors or radios or automotive controls or what have you—have had ideas as to what we could be attempting to exploit. We try very hard to support a very high majority of those things. Neither of those two factors would necessarily distinguish one company from another, except to the extent that the increment is truly different between this company and the other company. I would be inclined to guess that there has been a difference because we devote so much time to both of those factors. So we have the energy of many, and we have the continuing admonition: Are you sure you've told us everything that you want to tell us that you want to try? We still have to cut some things off at the bottom line.

We have an expression here that is illustrative of that, and it goes back many, many years to when we established a program called the Technology Road Map. That, too, is not so different from a lot of other people, except when we started it we didn't know of anybody else that had one like it. A road map is indeed a road map. But the road map ends up by putting, in effect, destinations and routes on it, technical destinations and routes. One of the things that we encouraged was that there would be a more than adequacy of minority reports. Once you and I have elected to support what was to be on the road map, then we'd go back and say, "Who didn't get heard?" Or "Who got lost from being identified?" Then that gave us another iteration; since maybe we had been too shortsighted. What I'm attempting to do with these factors that we're talking about here is to identify factors that are absolutely relevant to the engineering community.

Also we want to find some other more general factors. We've tried to practice high integrity with the customer, for example. These are also probably distinguishing factors if indeed any company does them incrementally differently or better. But those are two factors (WE, reach-out risk) that caused Motorola to have a pretty generous share of new products coming along or to be among the first to market these products. Then it's been within that environment that the professionals have done—compared to what we did in prior years or decades—superior jobs in identifying the spectra, specifying what they wanted to design and readying them for production. There is also the manner in which our scientists and engineers have teamed up better than they did before with all the other functions of the institution.

So, in summary, I think it's a case that we have done a rather good job of seeing to it that all the influentials had a firm voice in *what* we ought to be doing. We kept encouraging them to look for more, and letting them know that we could support them. Then they have put process alongside of intent, and they have performed quicker, better, moved earlier into production, provided better yields, all those things.

Aspray: This sounds like good advice for any kind of company. Whether you were running a service business or a product business, and whether it had to do with technology or not, those maxims that you've suggested would apply. What is it especially about being a technological business?

Galvin: I guess I would emphasize that I think what I have said was intended to be distinguishing of a technology company because we have tens of thousands of engineers, and they have technology anticipations and preferences. We search high and dry throughout our technical community to make sure that the best idea for a totally new function to be accomplished by our technology or a radical change in the improvement of that function is one that we are aggressively investing in in the laboratory.

Incidentally, we engage in probably as much self-development, teaching, of laboratory people as some in the business do, and a lot more than others in the business do, on the assumption that all engineers—I'm not an engineer, but those who would be—can get much better at what they are doing. We are not satisfied with our apparent rated ability to design, and we are sending our people to school and urging them to do all manner of things that have to do with improving their process of development.

We currently have in vogue an organized attempt to change the rate at which development programs will go from initiation to exploitation, and we are following all manner of particular processes that will cause three, or ten, or twenty engineers to be able to effect their work at twice the speed, or ten times the speed, depending on the nature of the class of work, compared to how they used to do things. We're looking at goals that used to take thirty months, and we now want to do them in three months. I think they are all doable in due course. So we are teaching ourselves to run laboratories with higher intensity of expectations of what we can aggregate, with the expectation that we can do a lot more things, and do each thing with much greater dispatch.

Aspray: You say that you yourself are not trained as an engineer, but what kind of knowledge do you have to have in senior management of technology to be able to make appropriate decisions?

Galvin: Let me tell you an anecdote. It will illustrate my point. I've received one compliment in my life—I don't receive very many and shouldn't—but one of our senior people, who is an engineer and who is moving toward retirement and has never had to worry about what he said to me, said, "You know, we're very lucky we've had you as the head man of this company for a number of decades." I said, "Okay, I know you've got an angle. What's your angle?" [laughter] He said, "We were lucky you weren't an engineer. If you were an engineer, you'd probably have known some of the reasons why we were pretty sure we couldn't do what was tossed out as a speculative thing. You kept believing we could do it, and then we'd have to go off and try to do it, and we'd discover we could do it."

I think what a person like myself can contribute, if one has had the experience (and I have had decades of experience making these judgments), is that one learns to measure the capacity of an institution or cadres of people in it. You learn to bet on good folk, and lots of companies have good folk. But I would bet a little stronger on many of our people than sometimes they would bet for themselves, and often enough they would prove their capability. All I was was an expresser of hope or faith in them, and trusted they would get it. Forgive all those apparently soft words. They are very powerful words. We deal with a sense of confidence and dependence on each other. If each one of us reaches out a little harder and finds that we do have that extra capacity, I think that's one role that someone like myself plays.

Aspray: How is technical knowledge represented at the senior level? Is there a particular person representing the laboratories, say, that is on a senior management team?

Galvin: In our company the overwhelming majority of people who are in leadership roles are engineering-trained. During the longest tenure of the chief executive office—while I was in that role, and I held it for thirty years—during all of that time, there was always an engineering-trained person in that office with me. For the last fifteen or so of those thirty years that I was the chief, two of them were engineers. They were superb engineers in their practicing days, and one of them until this day. He is still a vice chairman of the company although not in the chief's office. He was probably as acutely tuned to today's technology as even our younger people. He couldn't design the last micron of specification on an integrated circuit. But he would be able to evaluate substantially what the very bright young engineer may be explaining. So we have that engineering knowledge proliferate throughout our institution. When I was active in the direct management of the company, I would be rubbing shoulders with these people every day. So I picked up a lot of technology by osmosis, and then I did my own extrapolations from that in terms of my expressions of confidence. We did have to talk the language, and I'm able to talk the language and estimate what's going on, though I could never lead an engineering activity.

Aspray: For those people who come up through an engineering track, do you feel there's a value in either placing them in positions where they get business experience or getting special training on the business side? Business school or continuing education?

Galvin: The answer is yes. Organizations go through stages or eras, and during the first fifty years of Motorola's existence—we're sixty-five years of age—most of that was accomplished by experience. If you were a very bright young engineer, and your product line was doing well, we saw to it that you had considerable interface with customers. You discovered there were a few things that weren't strictly engineering that helped the engineering result to be translatable. You learned to become a business person as well as a technology person.

We've been transitioning into a new pattern. Today we recognize that as our growth rate, which continues at roughly 15 percent a year, begins to require very large absolute numbers of growth every year—15 percent of ten billion is 1.5 billion, so we're creating $1.5 billion worth of business every year, and pretty soon it'll be $2 billion worth of business every year—means more people have to accelerate their ability to have the general influence as well as the technical influence. So today we are moving towards having to substitute some experience with a scholastic effort of some kind: going to an outside school, doing a great deal of inside training, training on technical subjects as well as business subjects, and training through role modelship.

If you are a leader and a good role model, you alert staff to try to recognize their strengths and weaknesses and work on them through formal schooling. As a matter of fact, we are now having to move in a more formal fashion to accelerate people's attention to knowing the generals as well as the specifics. The specific is the technology. Our election in this regard is very simple—I think it's the same election almost any other company makes—that you can't take very many general people and cause them to be very good at technology. But you can take people well trained in technology, and they can become generalists. We have a great deal of confidence in many of our people being able to do that. But we try to be selective and not interfere with, let us say, your technical potential, if indeed that is even more rewarding to you or is more beneficial to the company. So to the extent that we can each play our selective role, mutually we try to decide which is best for both of us, the employee and the company. But if being a generalist is the destiny of an individual employee, we have to provide a great deal more formal training.

Aspray: Do you find that the kind of formal training that's available outside of the company in universities, say, is adequate to the needs of the company?

Galvin: Only partially. To the extent that the engineer has had very little experience in knowing the principles of marketing, there are excellent things that can be taught in that regard. If one were to wish to focus on the issue of international culture, many, many things are offered at the university level. But ironically, the most important class of subject to us is not available in the university, and that is all of the aspects of things that come under the rubric of quality. Universities never did anything—that's almost an accurate statement. There probably are three, four, five—counting on one hand—universities that have done some original work on the subject of quality prior to 1980 or 1985. And few had even done anything by 1990. So we had to create our own class of thoughts, organized ways of thinking and doing about quality, and have had to teach that to ourselves. We had to actually teach the vocational skills, and we had to teach the attitudinal judgment class of factors. That has been an immense void that we have had to deal with. I think we have dealt with it rather skillfully.

Speaking again to your first question, that would be, I think, a distinguishing feature of Motorola; that most companies have not gone to the extent that we have to sit ourselves down and say: "You are a factory manager; you must learn all these processes and procedures and concepts. You are an engineer; you must learn all these processes and procedures and concepts. And you must use them." They can be used in some reasonable self-directed way. But the principles are pretty straightforward and are objectively proven as being worthwhile.

Aspray: Every company today pays homage to quality and to customer service. But it seems from everything I've read about Motorola that it is different from many companies in that it really takes quality seriously, and has been able to implement this concept effectively in the company. Can you speak to this issue?

Galvin: Yes. I think your observation is fair—valid—and would not be exaggerated for us to confirm it to you. There probably are fifty or a hundred companies that have applied a devotion to quality. Just to mention a couple of them, I think the Milliken Company is certainly such a company. Xerox is very close to being that kind of a company. Then there are other companies that have been coming along in this area, such as Proctor & Gamble. Parts of AT&T have embraced it very substantially. Texas Instruments is coming along rather well. I'm sure I'm leaving out some that deserve to be mentioned, but we'd be hard pressed to carry this part of the conversation on very long.

We have placed an inordinate devotion in quality. The reason that we have succeeded, I believe, is that everybody who could be an influence, and that started with me but it continues with our present chiefs and lots of the other main players, have embraced the proposition that quality is the first item on every agenda. I don't mean that in the sense of writing down a ledger of subjects for a ten o'clock meeting. But it's *the* first issue. We hardly ever start anything without saying, "What about the quality?" We are even that pedantic about it, if necessary.

Aspray: I see.

Galvin: We'll say, "Let's talk quality first." As a consequence of that, all of the subpieces end up having to be addressed. So our people are addressing quality issues at all times: in design, in designing for manufacture, in being able to manufacture what's designed, and all those kinds of interrelated factors. There's no place in the company where it doesn't get superior attention, whether it's personnel or accounting or any other of the so-called soft subject functions of an institution. It is everywhere in the culture. If there was an engineering department that were to prefer not to be burdened by this, they would be surfeited by everybody around them—all of whom were practicing it inordinately.

I think we're also doing something that is helping to complement that a little better than a lot of other companies. We have long been a participative management-oriented institution, starting back in the late forties, and we've muddled along for a long time learning what it all meant. But the culture, the roots, kept getting deeper and deeper in the fifties, sixties, and seventies. Now the thing that the world refers to as "teaming" or any other words that are metaphors of that, these are things that are second nature to us, including very many engineering teamings. Our engineering teams, as a function of the things they will study, are coming up with seminal concepts of how to design, not what to design. Seminal concepts of new procedures to engage in in the laboratories. We are trading those ideas among a lot of our people. So with the culture of participation, teaming, and all the other relevant words that again are on the lips of a lot of people, it is ingrained here—it is a very natural practice among a gigantic majority of our people.

Aspray: You mentioned the soft parts of the company where you also wanted to implement that. It requires, in a way, being able to quantify—or at least codify. How did you go about doing that?

Galvin: Early on, in the generic objective of our having a quality program, our people learned or conceived things—I'm sure they mostly learned from somebody else to begin with, but then they augmented or incrementally conceived things that were better—that satisfied the objective that we were going to have to have very substantial metrics. The word *metrics* was not a commonly used word in Motorola in 1980. It was a very commonly used word in 1985. By metrics we meant, "measuring everything." In particular we measured in cases where something was not up to snuff, a defect. Anything that isn't right is a defect. We adopted the heresy that it was well to have a proliferation of data. This was in contrast to the common thesis that one should have the least data possible because data will just obfuscate things, and one does not want to have to be just keeping a lot of records anyway. We keep records on virtually everything around here so that we can use that data for first-cause analysis. We are constantly flushing up for ourselves that there seems to be a problem in a particular place. A yield-linked condition or a warranty issue, all these things that relate to something that wasn't adequately designed and produced.

Early on we said, "We need an inordinate amount of data." The engineers had to keep more data, the factory had to keep more data, which all went back to the engineers. The field had to keep more data. The accountants had to keep more data on what they were doing. Let me go off on a tangent for just a second. We keep prolific data with regard to the accuracy of our shipping tickets, our receiving tickets, the receivable documents, the bills we send to our customers, which is a receivable for us and a payable to them. If we ever find one defect, a number that's out of place and it's therefore difficult to trace the paperwork, that's a condition in Motorola that we consider intolerable.

What you have now is a culture. It's intolerable to have a mistake on a piece of paper because that means the poor customer has got to go around and check all kinds of records and figure out which two pieces am I supposed to compare before I pay or I do something else? It doesn't have any value. That could very well have been because we didn't put the label on the product right when we designed the product, and that's a soft part of the design, but it annoys the customer even more than if the product didn't work when he took it out of the box. At least they'd say, "Well, once in a while maybe products don't have to work. But, gee, can't you get the paperwork right?" If the culture requires that even the paperwork be right, sure as hell the engineer knows he's got to get the circuit to have a life cycle benefit to it. All of these things reinforce each other.

Aspray: Maybe this is a simplistic question, but in what ways does quality benefit the operation of the company?

Galvin: It is axiomatic that in a good company, one that is operating to a good set of standards, is accepted in the marketplace, and is getting a decent share of the market, et cetera, but which does not place a superior emphasis on learning the practices of good quality, the cost of inadequate quality to that institution is probably between 20 and 40 percent of sales. That is the penalty to that institution for lack of attention to superior quality—20 to 40 percent of sales. So if you have a company that is doing, let's say, $100 million worth of sales, it should be probably spending $20 to $40 million less to produce the same result. People say that's inconceivable. That is a fact. We know it's a fact by very sophisticated analysis, and anybody that's deep into this subject will confirm them. They may instead say it is 15 to 30 percent. But twenty is an easy number to deal with. If you can cut that in half by some very conscious efforts and in ways that are evident, one starts to say, "Well, that I can understand. Now you've saved me some money."

Most businesses, and even most engineers, have a propensity to actually put too damned much emphasis on analyzing what the money factors are. But if that's what gets your attention, we can use this one to get the engineers' attention or anyone else's attention in the company. The payoff from an accounting or a financial standpoint is gigantic. It's just gigantic. Before you leave, I'll give you a little sheet of paper that is quite public around here, and we give it to anybody, called the "Welcome Heresies of Quality." It is a one-pager. Have you seen or heard of that?

Aspray: No, I don't think so.

Galvin: One of those heresies is YOU CANNOT RAISE COST BY RAISING QUALITY. This is saving enough to the pragmatic, traditional, business person. There are, however, more valuable, more persuasive reasons for quality. They are measured in customer satisfaction.

I sat in the corner of a cafeteria in the Ford Motor Company about ten years ago. I had spent one whole day with whomever I could visit: people that installed our electronic equipment, who serviced it, who paid the bills, and so on. I didn't see any big shots. One fellow berated me for ten or fifteen minutes about the fact that he couldn't reconcile our shipping tickets and their receivables with the payables they had. This did not add any value to the Ford Motor Company. He spent three weeks trying to reconcile this. Ford Motor was dissatisfied with Motorola. This person represents Ford Motor Company, and he was telling everybody at lunch breaks and so on that "I don't like doing business with Motorola. They waste too much of my time." When we fixed that matter and finally he had months and months of satisfactory relationship, the big value now is we have not only eliminated the complaint, but we may have somebody saying, "Oh, I really like doing business with Motorola." Well, that's invaluable. There is the greatest value of all. Whatever is the way of articulating total customer satisfaction is the really great benefit.

How does quality affect companies like ours? There was a recession, we are told, in 1992 and 1991. It didn't impact Motorola. That's an inaccurate statement. So I'll put a parenthetical interpretation in there. Recessions impact everybody because you end up with price competition that's more exaggerated, and very vigorous efforts and plans on the part of competitors who are trying to hold their position. So we have to scrap harder to get what we get. But we kept increasing our share of the market because now there was plenty of supply, and it made sense for the guy who wanted something to buy it from the best supplier. If we're the best supplier, not only best, but our quality is best. Our prices don't go up because quality takes our costs down. So we get a bigger share, and we weren't laying people off in any of our businesses. Our sales were holding up, we had some growth. There was a little pressure on the profit margin, not because the quality was up, but because the other fellow was saying, "I'll practically give you my product." We had to meet a price competition. That's the American Way.

So there are gigantic payoffs, and the engineers appreciate that. It really pays off to the engineer in that, if his company's sales volume holds up, then there's the opportunity to continue to allocate whatever your percent for engineering is against that high base. So the budgets for engineering hang on. The reaction is, "Gee, I guess I'd better design better, design faster." We get to market faster, we get our market share up; we keep the sales volume up. The whole thing cycles around. You know that story very well. Our people understand that. So quality is absolutely the supreme driver as far as our company is concerned. Some people say, "Oh, yes. Quality's very good. We think highly of quality. But we think going for profit margin is very good, and we have a very special approach to advertising." Those are all good things, too. But none of them rank with quality in our estimation. In that sense we probably are rather unique in that we are biased overwhelmingly to quality being the determinant of total customer satisfaction, and that drives so many other things.

Aspray: How does it affect economies in manufacturing inside the company?

Galvin: Oh, gigantically. For example, if one has a perfect operation, you don't have to have any buffer stocks. So you don't have to have storage space in the factory. You don't have to have repair stations. As soon as you start to design a manufacturing operation where you don't even design space in the building for buffer stocks, you have a lower-cost factory. You don't have to have racks to put them in, so you don't have such leasehold improvements in your building. You don't provide any people for repairing the product because you have nothing to repair. You're going to probably have a more consistent demand from your customers, so you can now operate the factory a little more smoothly. Although nothing is ever totally predictable, you can tell your suppliers what their schedules are more intelligently. They say, "If you're a predictable customer, I can give you a little better price." So everything incrementally ends up being lower in cost. It's not just a case where somebody says, "Well, you've designed the product, and when it comes off the tool, this is what the material and the labor cost will be." It's all these other factors that make a 5 or 10 or 15 percent difference in every one thing that you do.

I'll tell you an anecdote. The Chrysler Corporation did a thing along this line when they conceived of the technology center. It's probably the most dramatic thing that American industry has seen in ten years. Have you been to the Chrysler Technology Center?

Aspray: No, I haven't.

Galvin: Do you know about it?

Aspray: No, I don't.

Galvin: The Chrysler Technology Center is a building, but it's also an activity, a function, and an organization in the Chrysler Corporation—built outside of Detroit fifteen or twenty miles. It's like a Pentagon. I don't think it's quite that big, but it's a big place. The Chrysler Technology Center is, in effect, a facility and an organization that permits for the total integration of the entire Chrysler Corporation under one roof. The people who think with the customer reside in that building. They go out and make trips. The people who conceive the product work are in that building. The people who manufacture the product work in that building. There's a practice manufacturing prototyping plant in it. Et cetera, et cetera. The whole business is integrated in this one great big building. Seven thousand people.

The building cost Chrysler Corporation one billion dollars. Lee Iacocca authorized that amount in 1983, and he was criticized until it was opened last year. I happened to be the one invited speaker to come to the grand opening of their facilities, and I knew what the gigantic promise of this thing was. When I got there, Lee couldn't come and hear my speech. He was upstairs, still defending this concept and expenditure with the press. He was challenged, "How could you possibly spend a billion dollars when you have all these other problems at Chrysler Corporation?"

When it was all over, I was escorted up to Lee's office. He was sitting there with his tie open and I said, "How are you?"

He said, "Oh, fine. I've had to handle the press again." I said, "Lee, I hope you were able to tell them what you and I both know is the case. The day that this facility goes on stream a hundred percent—and it's just about now in 1993 or in 1994—everybody will be clicking; the team will have been playing, and they know what they're doing—count 365 days from that point, and you will save in that 365 days the entire price of this project, one billion dollars. Not over 25 years, but in one year. Why? Because the Chrysler Corporation does $35 billion worth of business a year." At that point in time, the cost of inadequate quality may have been $7 billion, 20 percent. Okay. Maybe it was only $5 billion. But let's say 7 because the numbers are nice and easy to multiply. So I told Lee, "You will improve your operation by eliminating the cost of bad quality by at least one-seventh, 14 percent of the 7 billion the first year. And you couldn't do it if you didn't have this facility." So that's how it affects manufacturing, engineering, accounting, everything else.

Incidentally, it couldn't have happened without its being a design center, a technology center. It wasn't an accounting center. It wasn't a consulting business center. It must start out as a technology center because it's a technology business, making cars. We do something similar, but we do not have to do it the same way as Chrysler—because cars are so damned big. We can do something similar in pockets of our company. We have all kinds of little places like that, where everybody does everything in one room. The designs come faster; the manufacturing is lower cost.

Aspray: Is there a structural integration of design and manufacturing within Motorola?

Galvin: I guess the answer is "yes," but it's very diverse by individual pockets of businesses. There are group dynamic factors that make a gigantic difference, such as I just alluded to. Frankly, there are also disciplinary factors. In our case we have done the unique thing. Nobody else has done it to the extent that we have, namely, we have obliged each of our suppliers to promise to attempt to prepare themselves to be worthy to compete for the Malcolm Baldrige Quality Award.

That is making a vast difference with a large number of our suppliers, a lot of whom thought that was meddlesome, unnecessary, superficial, and superfluous. Once we put their nose to that grindstone and said you must be able to match the factors and criteria, the 33 of them, or however many there are, in the

Baldrige Award, and be worthy to compete for it, they discovered that they had to learn some new things. Then we teach them.

Aspray: I see.

Galvin: If they want to come, we'll put them in our classes, and we'll teach them what they need to know to have their quality system match our quality system. Not by rhetoric, but by practice. We will teach our competitors the same principles, on the assumption our competitors will be our suppliers. Because in this industry, as you know, we compete with each other, we supply to each other, and we very often consort with each other in some honorable ways. So we'll teach anybody to follow our quality standards and processes.

Aspray: That's very interesting. The Baldrige Award is not only a recognition of quality; it's a vehicle for obtaining it.

Galvin: Oh, yes. It should be the national standard, and we have advocated that the President declare that. Nobody's done it yet. Maybe this President will. We're dumbfounded to see that nobody else has obliged their suppliers to do this. IBM has partially done it. Maybe one or two others. How can they be *laissez-faire* with their suppliers if they want to be perfect themselves. If 50 percent of their parts come from suppliers, how can they be perfect if their suppliers are not put to this same challenge? As a matter of fact, by our doing this, we have given the challenge to the industry, because anybody that wants to be good supplier of the industry would like to supply to us. So it's good for everybody else in the electronics industry. In that sense, only one of us had to take that lead. But we get the greatest benefit out of it.

You are eliciting a few other particular reasons that relate to your first question as to "what have you done different?" I wouldn't have wanted to start by saying "Well, we forced all of our suppliers to prepare to go for the Baldrige award" and have somebody think that was the kingpin or pivotal factor, but you can see what the subpart tactics are and how they all relate to our emphasis that quality comes first.

Aspray: We started off on quality and customer service. We have talked mainly about quality. Are there some things that you want to say about customer service? I'm particularly interested in the fact that you're increasingly an international business, and in how you deal with your international base of customers?

Galvin: A customer is a customer. There is no difference between dealing with the Chinese or the Japanese or the French or the Germans or the what-have-yous. It's all the same. Our people practice these principles in various places of the world, incrementally better than each other. We keep ratcheting each other up. Every once in a while we discover the folks in Malaysia are doing something way better than the folks in Scotland, or the Scottish people teach our people in Austin, or vice versa. So the globality is of no moment, in terms of the objectives, or the standards, except that we find that some cultures adapt quicker, and then we all have to race to catch up to those cultures.

Aspray: But what about distribution of design, or distribution of manufacturing?

Galvin: It is pretty much all the same. For example, we have a manufacturing-research organization—I don't know if I'm using the right words here; it's a laboratory for manufacturing in two or three main places in the corporation. In every design organization, they're always designing for manufacturing. So we are teaching each other even from project to project. Whatever we discover is the best process, we'll put that into our Japanese plant, or we'll bring it from there and put it into our Phoenix plant, or put it into our East Kilbride plant.

Our orientation from this point forward is primarily to put factories somewhere else to serve markets. We are not putting them out for the sake of being the only or best answer to low-cost production. We are making a large number of our products now in towns near Chicago or in main towns in Florida, and shipping them all over the world. One of the great benefits of being in global markets is to be exposed to customers in parts of the world who may have different and better standards than customers at home.

Many of us have learned from the standards of Japanese customers. We decided a decade or more ago that we would find a way of totally satisfying Japanese customers and believed that if we could do that, we could satisfy any customer in the world. Today we're finding some other customers are also very demanding, so we learn how to serve them. Then we can serve anybody else's customers. That's one of the great benefits of having a diversity of customers, finding the toughest customer and electing that person to be the one you are going to satisfy. Then you make everything else easier.

Aspray: I understand that, especially in the mid-eighties, there was some concern about your Japanese competitors dumping and such. Can you describe the issue there?

Galvin: Yes. That issue is really very simple. Among the strategic factors in business—and one that is insufficiently appreciated by a lot of people—is that there are people who would attempt to practice the principle of sanctuary. The Japanese practiced it and continue to practice it. Before, very substantially. Today, partially. They try to keep their home market a sanctuary. They build their strengths in the sanctuary, like a military organization would do. Then from the strength of your sanctuary, you can go out and do mischief, which is some kind of aggressive, competitive phenomenon. Dumping could be one of those things. Then when they have finally done enough harm to their competitor in the new market, they can take their strengths and build up the Japanese strength in the new market. The Japanese recognize this. It's been a practice in the military for hundreds of years. And they practice it in business. So have some other countries. We recognize that and essentially said, "That cannot be tolerated because if they stay strong there, they will finally weaken us here, and we would finally go out of business." So we recognized that we had to break down that barrier. It was life or death. In military terms, you would have to attack the sanctuary of your enemy. We used every legitimate process that we could, starting with "Let's be sure we have the right product, on time, the right quality, so that if they ever do give us an order, our product will satisfy them from an operational standpoint."

Then we did all kinds of things to make sure that we gained access to their market, the opportunity to quote and be received with some objectivity. To do that, we had to use whatever legitimate governmental tactics were at our disposal, such as saying: "You can't bring that mischievously priced product from your sanctuary into ours and do harm. You're dumping, and we won't let you do that." We used that and many other tactics to open up the market. We used our government, which is, in our estimation, a very legitimate thing. People can call it "protectionism" if they want; it isn't. They can call it "managed trade" if they want, and maybe it is. But you've got to *manage* not to let your opponent have a sanctuary. So we were very aggressive in that regard. We were *the* leading institution in the electronics industry. We opened the electronics market in Japan.

As a consequence, it's been very rewarding to us. Except for in what was honest-to-goodness recession in Japan last year, we kept our share up, and it was very hard to compete and make a profit last year by anybody in Japan. The Japanese lost a lot of money themselves serving themselves. But we're there. They know that we're going to stay there, and we're going to grow there. Therefore, they don't have a sanctuary any more. That's how you achieve a level playing field, by making sure the other fellow doesn't have a sanctuary, and you can be there.

Aspray: Do you want to say any more at a philosophical level about the role of government in preparing an even playing field?

Galvin: First off, I think the private sector can and should do virtually everything for itself. But there are things having to do with policy, such as we've just alluded to, that need government attention; and I won't repeat any interpretation of that. There are a few recipes in technology that look like they do have to have some degree of governmental support. I'll start with something as esoteric as the fundamental understandings of the essence of nature—the supercollider for example. To me the supercollider and the Fermi Lab, et cetera, are very relevant to today's engineers. A lot of people don't want to accept that because you won't know how to peel back that subatomic structure until the year 2009, so why don't we wait?

Well, we won't get it by 2009 if we don't spend now. So we need governmental support for things of a fundamental scientific nature. We need the National Science Foundation to support the research base of our country, which is the universities. That wouldn't get done if there wasn't a facilitating agent, and there is no facilitating agent except the United States government, mainly the National Science Foundation but also a couple of others. We have the government labs, as a coincidence because of our atomic energy program, for defense purposes. As long as they exist, they're a marvelous facility that can be converted and are being converted to handling major increments of advanced research projects. I'm the chairman of Sematech, and I see where the collaborative activities of both the private sector and the complementary support from DOD [Department of Defense] have made a gigantic difference in that section of the semiconductor industry that makes the tools or the equipment. It looks like there is lots more science to be understood in semiconductors, so there is a place, I think, for the continuation of that joint research, although we may change the focus of one project versus another over the next five or ten years.

So I think there's a place for the government to do that. I think it's minimal in the sense that it will always be a minority of what the people will be doing. But I think we can find big slices of research that are of a fundamental nature, which if we can as a society know those principles very well, then all of the Intels and the Texas Instruments and the Hewlett Packards can go and do a great job of taking that scientific basic knowledge and converting it into technology and useful applied things. So there's a role for government support of fundamental research. I wouldn't want to see any company subsidized, but I think we can identify potentially winning—and here we are going to be selecting winning and losing ideas—ideas that look like they are worthy of pursuit. We can follow them through to the point of showing they aren't any good, which is also valuable knowledge, or finish them out and be able to transfer that science to the appliers. So I think there's a role for government.

Aspray: Two last questions. The first is about investment in tools and equipment. In the semiconductor business, every time you go down a factor of ten in scale of device, your costs shoot up a factor of a thousand, or something like that. How do you face this in the future?

Galvin: You face it by earning enough today to be able to afford it tomorrow. You can't earn enough today if you haven't anticipated and committed to the products that you can have early to market and earn enough on. Therefore, the cycle is a circle. We've been wringing our hands over this for forty years. I can remember in the fifties admonishing the guy that was running our semiconductor business: "Can't you learn how to run the semiconductor business with less physical capital?" A perfectly reasonable philosophical question to ask. But in the final analysis it was a naive question, and it had to be reshaped as, "How can we use the capital we're going to get more effectively? Because it's going to cost a lot of money to cook these silicon cookies." So we have just accepted the fact that it's going to be a very high physical capital cost.

When we have to buy a stiepper or something else, we want to figure out with quality systems how to get incrementally more out of our investment. If we can do that, and if the world is going to need whatever these circuits are going to be designed for, then we can compete. We'll make enough on that and be able to depreciate them enough. You will have to have some reasonable federal policies with regard to depreciation and allowabilities to earn some money, not be too heavily taxed on it—so that you can afford it. As far as we're concerned, we're going to be able to afford the next wafer processing plant. When it goes to the next level of investment, we'll figure out a way to afford it.

Aspray: I'm really very surprised by that answer. I had half expected to hear you say that it would be through collaboration with other organizations, through consortia and such.

Galvin: I thank you for adding that. I guess I would have hoped that I would have said that more clearly when I spoke briefly about the government involvement. But not in terms of financing our capital structure. No, I don't see that. If indeed there is a science need that needs to be clarified or created, that I think is going to benefit from—and we should even be obliged to use—some consortia or government support to reach new possibilities. But once the science is adequately understood, then hopefully the general policies of allowability for depreciation, tax policies, et cetera, will let the private sector price and profit adequately to put enough on the bones to where we can afford the next option. Motorola and Intel and TI are all going to have to be able to afford most of these things themselves.

Now, are there exceptions? Sure. We'll sometimes join others. We've got a small joint effort with Toshiba. It's a big thing in some people's minds, but comparatively it's small. TI has got a new plant they went into in Singapore with a couple of companies. You'll have little pockets of these kinds of activities. But in the final analysis, if TI were to have three projects with three teammates, that means they have the equivalent of one big project. So they've got to afford some very big capital investments, and so do we. So does Intel, so does Hewlett Packard, so does IBM, and all the computer guys, and so on.

Aspray: The final question is about the military and its relationship to industry. The military has been so closely tied to high technology development since World War II. I know your company has jointly developed a chip with the Department of Defense, and that a small but significant part of your business is in the military area. Has this military-industrial relationship been beneficial over the past few decades? And how do you see it changing?

Galvin: It has been beneficial, not in the direct transferability of one circuit over from the military to the consumer or commercial marketplace. But there has been considerable relevance. And, yes, we have been a part of military programs. We are absolutely a part of the Sematech. DOD-DARPA support things there. What we have done is to put limits on the nature of the things that we would offer as a service to the government. We never elected to go into the making of a missile, for example. But if there were missiles that were going to use some very new technologies, we will become a supplier of that subpart. We've found that that was a place we had value to add and a place where we could generate byproduct competencies.

You see, probably the grandest manifestation of this commercial payoff of military work was that our people had to become very proficient at understanding how to communicate through satellites. That derived for a few of our people a concept of a cellular telephone system with low earth orbital satellites as the way stations. But there's a derived competence to Motorola. Our people were able to extrapolate this idea from the projects we had. We've had a moderate number of things of that nature that have come along. That particular one, iridium, will be spectacular.

Now, what is the future? We think that our government—probably in your lifetime—is going to have to have a very extensive technology arms policy. A policy of having advanced technology applied to the armament needs of the defense entity. If we can maintain a competence, we will get a slice of that business. We will get the off-load or an off-benefit experience as well. So we intend to stay in that business. We think it's an honorable thing to support our government if they need armaments in that technology capacity. And we will get some secondary benefits from it. But we don't depend on it. I would say that would be a 5 to 10 percent factor of our business. The size of the defense business has always been less than 10 percent in our company. But it's another one of these great mountains to be climbing. If we can serve that market, we probably can transfer some competence over elsewhere.

Chapter 10

Mitchell Kapor

About Lotus

Lotus Corporation was founded in April 1982 when thirty-two-year-old Mitchell Kapor pulled together the financial resources to design and market the most successful software package in computer history, Lotus 1-2-3. He committed $2.7 million he earned by designing business programs for VisiCorp and $4 million from Sevin-Rosen, a new venture capital firm, to the project. Believing that the users of personal computers wanted the ability to take the output from one program and pass it on to another, he enlisted the programming skills of Jonathan Sachs to develop Lotus 1-2-3.

The program was an instant success. It offered a fast and flexible spreadsheet with graphics and database management functions. It did for the IBM PC what VisiCalc had done for the Apple II. IBM PC sales skyrocketed concurrently with the $495 Lotus program. In its first full year of operation, 1983, Lotus Corporation reported $53 million in sales. In October, the company went public, raising an additional $34 million. A year later sales had reached $157 million.

In no time at all, Lotus Corporation was transformed from one of countless start-up computer companies to a leading vendor of microcomputer business applications and, ultimately, to the world's second largest personal computer software vendor. By 1990 Lotus' customer base reached five million. Kapor chose to leave the organization in 1986, leaving to others the problems of running a company experiencing extremely rapid growth.

While continuing to be successful, Lotus was unable to match 1-2-3's tremendous success with its subsequent personal computer software packages. The company decided to branch off into software for larger computers and computer networks. Management also tried to diversify into other computer markets. An attempted merger with Novell to gain its networking business failed, but Lotus still managed a successful entry into the word processing software business. In recent years the company has developed its own networking products, including electronic mail forwarding software and the Notes package designed to permit groups of computer users to collaborate from different locations.

Mitchell Kapor

Place: Cambridge, Massachusetts

Date: May 20, 1993

Goldstein: Were there any specific experiences in your background that you found invaluable in preparing you for senior management?

Kapor: Let me start by saying I was not very well prepared for the management challenges that resulted from the very rapid growth of Lotus. This is managing in a start-up situation of hypergrowth. In our first year of operations we were a $50-million company, and in the next year we were a $150-million company. Nothing really prepared me for that.

But what was helpful—if anything was helpful—was that I had had a wide range of life experiences before this happened, as a counselor in the psychiatric unit of a hospital, as a disc jockey, and as a meditation teacher. For instance, I have a master's degree in counseling psychology. Some of the communications and group skills that I developed in the course of my graduate school and work experience in psychology turned out to be very transferable to the business environment. Some of my sense about how to motivate and work with small groups also came out of that experience. So that was useful. I also went to business school and dropped out of it. That was mildly useful.

Goldstein: Were you aware that your role in Lotus was shifting? When you were developing the product, were you conscious of a change in your role to more management-oriented responsibilities?

Kapor: I was conscious of an enormous change in my role due to the changing circumstances of the company. The nature of the management challenge shifted a lot from a very small, tightly focused group of people to a much larger, multilevel, and more complex organization. I consider both of those activities to be management. Some people might say that management only starts when you have a big firm, and that whatever it is you do when you're running things with only three people isn't management. But that seems to me to be an arbitrary distinction.

Goldstein: Was your technical familiarity with the products influential in your managing?

Kapor: Oh, completely. I think managers who don't have a firsthand knowledge of the substance of the products that they're making are at an enormous and usually fatal disadvantage. Having firsthand knowledge in the software business means having a technical background and fluency with software. In the start-up mode, if you don't know it yourself, who can you count on? Even in later stages of growth, if you are the CEO and you don't really have the technical background, you are forced to rely on other people. It's just very difficult if you don't know which questions to ask. Mistakes get made that could be really crucial.

Goldstein: Where did you find that expertise particularly valuable?

Kapor: Obviously in development it's completely critical. I would even argue that in marketing it's very critical. One of the things you have to take into account when you look at this issue of a technologist as manager is the phase in the industry life cycle of the participants. In the personal computer software industry at the time Lotus was starting and when I was running it, it was a very young and immature industry. The basic ground rules regarding how you went about doing development and how you went about marketing were not well established. In fact, they were being invented. There was a lot of trial and error.

Today it's very different, in that the basic ground rules are extremely clear to everybody because anybody who hasn't followed those ground rules is out of business. There is a different kind of expertise that is needed in leading this kind of company. Back then, you really needed to be very creative in the sense of discovering how to make products, what to put in them and what to leave out, and how to market them. That's pretty well understood today. But there is a fiercely competitive battle for the hearts and minds of customers. The product categories are stable, but the competition is much, much stronger than it used to be.

There is a different kind of expertise that's required to lead that. There is also far more complexity if you are running a company. Today Lotus is on the order of a billion dollars a year. The complexity of coordinating all that with these huge product lines and international operations and thousands of people in the organization is daunting. There is a whole set of management issues in the coordination of complexity that simply didn't exist then, or just existed in a very embryonic form. You need to be a specialist in doing that. Being a technologist isn't particularly helpful. I don't think it's necessarily unhelpful, but it's not an expertise I have in managing this kind of complexity. In fact, I don't like it because, for the most part, it is not hands-on.

Goldstein: Do you think it's more traditional business school type work?

Kapor: Yes, sort of. It's the sort of thing that business schools purport to teach people how to do. I don't think they actually do. The point is that my experience at Lotus was in the early days, at which time being a technologist was very important—in fact, critical—because of this need to figure out the ground rules. Without having the necessary technical knowledge, it's inconceivable that you could do that in the start-up phase of an industry.

Goldstein: The management issues are different in a mature industry?

Kapor: In a start-up versus a mature industry, yes, they are different. Or in a mature firm in a mature industry. You have two continua of maturity. One is the firm itself, and the other is the industry. You could have, for instance, a young firm in an old industry. The one thing you can't have is an old firm in a young industry. Actually you could if the firm was in some other industry, and they were getting into something new. So it's what I would describe as a two-by-two matrix of the age of the firm and the age of the industry. The interaction between being a technologist and being a manager is really contextually dependent on which quadrant you are in.

Goldstein: Do you have examples? Is there a disadvantage to not being a technologist in these first quadrants?

Kapor: Yes. There are a number of key decisions that we had to make, for instance, about the development environment. One place where many firms went wrong was not so much that they had the wrong product idea, but they went about developing it in the wrong way. They developed it in Pascal when they should have developed it in assembly language. Or they developed it on minicomputers and cross assembled it, whereas they should have developed it more directly on the target platform.

There was a certain wisdom then about good software development and methodology that would have led one to adopt the wrong strategy. In fact, a VP of development might well have suggested exactly the wrong strategy. So when the CEO is a technologist and also understands the business objectives and needs very clearly, then there is a better likelihood of making the proper decision, which in our case was to develop in assembly language right on the target machine for the IBM PC. What was at a premium was performance. That was the way to get the highest degree of performance. It was more difficult to develop that way than in higher-level language, but the performance sacrifices would just not have been acceptable to the marketplace.

A pure technologist might not have understood that it wouldn't be acceptable to the marketplace, and a pure manager might not have understood the magnitude of the trade-off. Whereas if it's one and the same person making that decision, you integrate those two facts or those two constraints together, and say, "Well, we've got to develop in assembly language."

Goldstein: But what is it that led you to that successful perspective?

Kapor: I knew that performance was paramount in terms of the delivered benefit to users, and I knew enough to know that the way that you would get the maximum performance is by developing in assembly language. In fact, we had been developing the prototype in C and we could just see that the performance wasn't there. This decision incorporated my beliefs, my values, and my observations about what users wanted.

Goldstein: Were there any special circumstances critical to the launching of 1-2-3 versus the introduction of the PC?

Kapor: Yes. I should explain the basic story: Jonathan Sachs and I were playing around with different ideas for an advanced spreadsheet, a graphic spreadsheet that incorporated other functions and had programmability to it. What crystallized our thinking was the announcement of the IBM PC. We had started before it was announced, and in fact we didn't even know it was coming.

It was apparent to me that that was the proper target machine for our efforts because of two factors: One was a marketing factor that IBM was going to be a big deal for business. Two is that it had certain desirable characteristics. The first is that they were going to distribute in Sears and Computerland computer stores, not through the IBM sales force. They had gone to the outside for their processor, operating system, and applications, which I thought was a very smart move. I feel they were violating their own conventional wisdom about how to do products. So I said there must be some very smart people at IBM because they already understand you have to take a different approach to succeed in this market. That, combined with the fact that it was IBM, which would make it very acceptable to business, and combined with the fact that it was a 16-bit processor, whereas the competitors such as the Apple II and the Tandy machine were 8-bit machines, made a big difference. I could see that users wanted to have more memory for their spreadsheets because they were running out of room on the Apple II. They wanted things to be faster so that they wouldn't have to sit there and recalculate. It was clear it was the right platform.

What happened is that there were two spreadsheets available for the IBM PC from the beginning. But both of them were basically warmed-over 8-bit versions. They only supported a total memory size of 64K, whereas with the IBM PC you could have 640K. I happen to know that they weren't optimized for the IBM PC. So I had to place the bet that if you came in and did a program that was optimized for the IBM PC—to take full advantage of its capabilities—and if you added the graphics to the spreadsheet and put in the user programmability and improved the interfacing, that's a winning combination. That was the hypothesis that we set out to test. In fact, that's actually what we did, and it worked.

We also raised what for that time was a significant amount of extra capital—initially a million dollars, and then later some more money—in order to do a big marketing campaign and introduction ads in *The Wall Street Journal*. That seemed to me to be common sense. If you want to meet the business market, don't advertise in *Byte Magazine*, instead advertise in *The Wall Street Journal*. In fact, it was common sense. We just happened to be among the first people to figure out what was sensible and go do a good job executing it.

Goldstein: If that was common sense, were there any early business decisions that you made that were less certain?

Kapor: Well, of course there were a lot of people who thought the whole PC industry was a big shot in the dark.

Goldstein: Did everything seem crystal clear to you?

Kapor: No. The thing that seemed to be really frightening was that VisiCalc had an iron lock on the spreadsheet market on the Apple II, and there they were on the IBM PC, and it was selling. MultiPlan was also in this. Both of those products were distributed by IBM, as well as by their vendors, VisiCorp and Microsoft. So the idea of having yet another entry in that category seemed to me quite frightening. I said, "People are not going to switch." At least I was afraid that they were not going to switch. We went to a lot of trouble to try to position 1-2-3 as integrated software, not as a spreadsheet. That lasted about a month after we went out with it, because it was clear that it was an enhanced or second-generation spreadsheet. So we let the market reposition us.

One of the peculiar characteristics of the whole Lotus story was how quickly things happened. There wasn't a lot of time in which there was much uncertainty. In other words, I spent about three hundred thousand dollars of my own money from royalties for a product that I developed for the Apple II. I was running out of money, and I went to try to raise venture capital. I was able to raise it almost immediately. In the context of doing that I wrote a business plan which was ludicrously far off in the revenue aspects. We thought we would do three to four million dollars the first year. But it said we are going to deliver this product, and we are going to do professional marketing, and we are going to run with it. And that's what we did.

We actually got the money in April 1982, and we announced the product in either September or October. We went to Comdex in November, and we wrote a million dollars in orders on the floor of Comdex the first time we had gone out. We knew that our estimate of three million for the whole year was ludicrously low, so that's when we started raising the forecasts. We shipped the product on time in January of 1983 as promised, and then we had this $53-million-dollar year. The major challenge, actually, was in implementation. The one thing that we did very well was that we executed. We met all of our commitments and our deadlines, and we shipped. We built a factory that was able to meet demand. I don't know how many units we sold the first year. But I can figure it out. What is 50 million divided by 250?

Goldstein: Two hundred thousand.

Kapor: Yes. Two hundred thousand. We also had manuals to print. The biggest challenge was the implementation challenge of having to scale up. We had to build everything from scratch, including a whole financial organization capable of registering the revenue and paying out. We hired hundreds of people so there was a human resources function. Then we had our marketing. We were training dealers; we had road shows where people would go out and train dealers. We had customer support organizations. We had everything that you find in a modern personal computer software company. We built it all on the fly during that first year.

Goldstein: It sounds like it takes a lot of savvy to know where to get the venture capital, to go deal with Comdex, and build a financial organization.

Kapor: I had had four years of very intensive internship in the personal computer software business. I had been a consultant. I had run a cottage industry software company, published a lot of stuff, and sold it mail order and through distribution. I worked for a Silicon Valley start-up. I had started out as a product manager, which is where I learned a little bit about venture capital. I had as much experience as anybody in the business side of it. There were a lot of people who had worked for Digital for ten years, or Hewlett Packard. But it turned out that most of what they knew was wrong. It was unhelpful in this new industry with new rules.

Goldstein: When you say wrong, do you mean inapplicable to the new industry?

Kapor: Yes, inapplicable. I think I got a little lucky with the venture capital. I knew Ben Rosen slightly, and I went to him first. I said, "I'm thinking about doing this thing." So he said, "Write a business plan." I wrote a business plan, and they decided to

	fund it. I had a track record. I had written two products myself: Visiplot and Visitrend, both personal software.
Goldstein:	That's where you earned the three hundred thousand?
Kapor:	Yes. That's right. They were companion products to VisiCalc. In addition to having gone to work for that publisher as a product manager, I had also had this other relationship as an author. They had the publishing contract. It was a tiny industry and I knew all of the principals in it. Anyway, Ben Rosen and his partner L. J. Fevin decided to take a chance on me. In conventional terms, if you looked at my resume, I had no management experience to speak of. Another thing is I had no technical experience. I don't have a degree in computer science. I had a couple of courses. I am a very mediocre programmer. The programs that I wrote and that made money were in Basic because I can program in Basic. I can't program in assembly language well enough to do a commercial product. So viewed that way, why would anybody invest? On the other hand, I had a lot of experience, as much experience as anybody, and am self-taught in terms of being handy around computer work. They decided to take a chance. More power to them.

The rest of what we did, in terms of the launch and the strategy and so on, was just applied common sense. I sat and thought about it. There was a team of people who contributed to it, and we had to make a series of decisions. I don't remember us agonizing over this. In fact, it seemed kind of obvious. It was scary because we didn't know if it was going to work. It was a lot of money. But we just had the foresight and the good fortune to be in the right place at the right time.

Over the years literally hundreds of people have come to me and said, "I have an idea for the next Lotus 1-2-3." I find this pretty amusing because I try to explain to them that we didn't know we were going to be the next Lotus 1-2-3. The idea didn't exist. We had something much more modest in mind. What we did in that kind of hypergrowth, the circumstances have to be very special and appropriate to it. You have to be in at the very beginning. It has to be far enough along that the market has potential for explosive growth, but not so far along that somebody else has gotten it. You are either there or you are not there. It's one thing to say, "We were smart; we recognized there was an opportunity." That's true. But I was fortunate to be at a point in my life where I was able to start a company right near the beginning. Not everybody can do that.

Goldstein: One thing you didn't mention before was pricing strategy. How did Lotus's price compare to VisiCalc?

Kapor: It was a lot higher. I forget what VisiCalc was, whether it was $195 list or $295. We went out at $495.

Goldstein: Was there much precedence for software that expensive?

Kapor: Well, D-BASE had started at $795 and might have dropped by that point. I basically wanted to go with the highest price that we thought was supportable. Believe me, we didn't run focus groups even on this. $495 just felt like the right price. It turned out that it was okay. There's another thing. A fully loaded machine, the most expensive machine that you could buy, was about $5,000. A minimal machine was about $2500. So at $495 it was 10 percent of the fully loaded machine or 20 percent of the lesser machine. That seemed to make sense as a ratio.

Goldstein: Were there other decisions about site licenses?

Kapor: We didn't have any site licenses at the start. That came later. International distribution also came later. We decided to go with retail distribution. We did decide to start a corporate sales force to call on corporate accounts directly, to stimulate volume purchases. That just made sense. That's where the business was. That's what the customers wanted. But we did fulfillment through the retail channel for a long time before we figured out how to sell direct.

Goldstein: You alluded to the success of IBM. One factor was the IBM name. That suggests that you believe that brand is an important asset in this industry.

Kapor: Yes.

Goldstein: How did you use Lotus's brand name once that became established?

Kapor: I think our conscious efforts to exploit it were much less successful than the momentum created by the success of the product. People identified the use of the entire machine with the use of Lotus as a spreadsheet. They saw the machine just as a vehicle for doing spreadsheets, which benefited us enormously.

Goldstein: Yes. The old PCs that have the Lotus part burnt into them and that's all.

Kapor: Yes, that's right. The market ordered us because they liked the product. The product was extraordinarily useful. It empowered a whole class of people in business who were nontechnical professional people. It gave them a productivity tool for this extraordinarily wide range of uses, anything involving calculations with numbers—not just financial ones. I | was a godsend. The product fully exploited the capabilities of the original IBM PC. We did such a good job with that. You can argue that this feature maybe was a little clumsy or we left something out, but nobody could do a better product because we had just hit the spot. Now, of course, Windows is another deal. But between 1983 and the late 1980s you just couldn't do any better. Between the momentum from being the category leader and the product excellence, it was a self-sustaining phenomenon. When we marketed other products, even though they said Lotus, if they weren't right, they didn't sell.

Goldstein: I'm curious about that expansion with Jazz and Symphony. Did those seem like further exploitations of the enhanceabilities of PCs?

Kapor: That was the idea with Symphony, but it was the wrong approach. It was based on miscalculation, a misunderstanding. We understood that there was a market for different productivity tools, spreadsheet, word processor, database and a market for integrated software. But the integrated software market turned out to be at the low end for people who don't want full functionality. They want a little of this and a little of that in a single, simple package. We didn't do that. Symphony was an integrated package that tried to be all things to everybody, and it failed by that standard. It failed to create a huge amount of momentum. Over the years it has generated hundreds of millions in revenue, but it didn't have this kind of momentum to propel the company. We misunderstood what it was that the market wanted. This was not surprising, given that it wasn't clear to anybody. Technically speaking, Jazz just did not work well enough. It did not do enough. Its spreadsheet and the other features in it were not good enough. Jazz was Symphony for the Mac. It had the same problems of product concept, and it had additional problems of performance reliability. Microsoft took the 1-2-3 lesson. When they did Excel, they did an exceptional spreadsheet for the Mac that was tuned to the Mac as a platform. They understood that that was what they needed to do. When they did that and got out there, they owned that category. And they still own it on the Mac.

Goldstein: So you are saying that an understanding of the hardware is indispensable.

Kapor: An understanding of the capabilities of the platform. Not so much just the bits and bytes of the hardware, but what you can do with it. That is absolutely critical. Also understanding what the market wants. People just want the most kick-ass spreadsheet they can get for the money that takes best advantage of whatever platform it is they are buying. So the Mac had a graph phase, and the PC didn't.

Goldstein: Would you play to the hardware in appreciation of the power of the hardware above the quality of the software?

Kapor: No, I wouldn't. Now we're talking about design. We're saying, "How do you design a successful product?" Appreciation of the platform is clearly one thing. Having to write a specific feature set is clearly critical.

Goldstein: How about on the execution? You know, the code itself?

Kapor: That has to be first rate also. You can have good ideas on the right platform, and if you do a lousy job in implementation—meaning that it's too slow or it's too buggy or it takes too long to come to market—you will get killed.

Goldstein: So it all has to be there.

Kapor: It all has to be there.

Goldstein: You were talking about the value of the technical background for managers. I was wondering if you brought anyone up in Lotus who didn't have that, who hadn't put in time in the technical area, who was a product of a business school or management experience. Was there ever a time in Lotus's history where you needed people like that?

Kapor: I have to characterize what happened a little bit differently. We were in something like a war zone. Usually you think of tragedies, like natural disasters, shipwrecks, or things like that—things that produce great unhappiness, as being the source of huge amounts of stress for the people that are involved. But it's also true that hypergrowth and enormous success produce exactly the same types of stress, without being life threatening, of course. We had the problem of building an organization and building a management team from scratch in a big hurry. I didn't know anything about how to do those things. I had a very mixed record. I built a management team of people that worked pretty well up to a hundred million dollars a year. But we were that size for only eighteen months. The job outgrew everybody. My job probably outgrew me, too.

Finding people to do finance and manufacturing actually was not so much of a problem. While the challenges of a hypergrowth environment were real, the financial functions were more or less recognizably the same as a financial function at some other part of high tech, but not PC softwares. You could take somebody who was good, and they could probably adapt. But marketing and development, key functions, were very difficult. We had a real lack of success in finding people who could do that well, hiring them and moving them along through the ranks. We tried, but it just didn't work.

Goldstein: Would the results speak to the quality of the person? Or was it more a question of whether their common sense coincided with your own sensibilities about these matters?

Kapor: More the latter. Part of the problem is that if I were doing things again, I might do them quite differently based on what I've learned. My management philosophy, for better or worse, suggests there needs to be a coherence of vision among the senior management. The CEO sets the tone and provides the overall leadership. If it is the wrong leadership, then the board has to fire the CEO. So I would look not for uniformity of opinion, but for the sharing of a sensibility in which there is a lot of individual differences of opinion as people approach subjects, both from their disciplinary specialty and from their own personality. I really think that that is paramount. You have to have the skills within the particular discipline. This was another problem since what it meant to do good marketing of PC software was so unclear. You had this additional problem of how to find people. How do you know what you are looking for once they come? How do you tell? We had those problems also. Now it's somewhat simpler.

But there is a question lurking in this regarding not only one's management philosophy, which I've given you, but one's philosophy of business. What's the point of being in business? People in business generally don't have either the time or, for that matter, the disposition to sit around like a bunch of Socratic philosophers and ask themselves that question. That's part of what makes people business people. In fact, it's one of the things that creates some tension when I try to fill that role because I'm naturally reflective about these sorts of things. I have an interest at heart in creating great products. The business was a kind of vehicle for bringing great products into the world. We were trying to do that. That turns out not to be a great business philosophy because you succeed in business by building the business, not by doing great products. There is a linkage between the two, but often the commitment to do great products and push the edge of the envelope actually interferes with business success in a number of ways that I can see in retrospect.

When you ask what's important, there were things that were important to me then that are still important to me now, but are somewhat at odds with the answers I would give if you were to say, "What should somebody do if they really wanted to build the business?" I would say, "Do what Microsoft does. Don't worry so much about product elegance. Worry about building." The first and foremost question is: How do you build an invulnerable grip in all of the key segments of the marketplace? It may turn out that product quality has something to do with it, but that could be about third on the list. It could also be making the right alliances, finding time to market independent of product quality. I just don't think like that. Or, I didn't think like that. I don't know what I would do now. I wouldn't run a company like that because I don't like facing those sorts of choices.

Goldstein: You talked before about the importance of coherence of vision. I wonder if different cultures developed within Lotus, if you had people there from the beginning versus other people who had different management skills.

Kapor: It's not quite that simple. But you are on to something. Since we hired so rapidly, we never had a single culture. We had fragments of many different cultures because we hired a bunch of people from Company A and a bunch of people from Company B. I tended to hire people that I felt comfortable with, which meant that there were a higher proportion of people that were a bit off the beaten track in terms of their life experience. Those people very quickly left or became marginalized as the company grew, and that's too bad. I feel badly about that still. There wasn't a tension between the old and the new. There were quite a few people that didn't have the same inherent love for the substance of the business that I did. Some of the other people did early on, and they came to occupy more dominant positions inside the company. It turned into more of a hard-headed, aggressive type of operation.

Goldstein: You were talking about the different things that are important to a business's success. Product quality may not be number one.

Kapor: No. By the way, I have to say something to that. When I was doing things, I had a fanatical orientation towards issues of product design and quality, end-user experience. It's not exactly fair to say that other firms that I'm implicitly criticizing don't care about product quality. I think they do, but in a more measured way of finding the right set of trade-offs between creating the user experience in the product and other business objectives. I do not want to imply that I thought that other firms besides Lotus do not care about product quality. I just have this fanaticism about it.

Goldstein: I wonder where you would position product support in that list, and how your attitudes about product support developed over your time at Lotus?

Kapor: You need to sell a successful user experience, and some degree of product support is critically necessary. The landscape of product support has really changed as we have moved from a totally uneducated user base to a user base in which there are a very large number of sophisticated users. This is another one of these issues where you need to know where you are in the industry life cycle.

When we started out, there was not a lot of the infrastructure or user education and support that exists today. Today there are hundreds of books. The best books in the category are much better than the documentation supplied by the company. There are third-party training companies. Large corporations do all their own training, support, installation. We had to do a lot of things because there was nobody else to do them back then. We did them: We trained the dealers, and we would answer anybody's question about anything. At a certain point the demand for support outstripped our ability to provide it. So the phone lines would be busy. That situation was eventually ameliorated as users got more knowledgeable and people started supporting themselves, and then third-party support of various kinds kicked in.

Goldstein: Were you ever in a position where you had to make one of those trade-offs between support and enhancement of product?

Kapor: What happens is you build up a big installed base, and you become a prisoner of it. From a business point of view you can't add new features that enhance the power of a product just because they're great ones. Users don't want it. They say: "This is too much. We are not prepared to accept things at this pace." Or, "We don't care. We are only interested in the six basic things." So there is a lot of tension there. I didn't like that.

Goldstein: I read in some interview you did in 1985 that you had just begun divisionalizing Lotus. I'm wondering what was behind that move.

Kapor: We need to have the history on the record. In mid-1983 I hired Jim Manzi as director of corporate marketing. He became director of marketing in the fall of 1984. He became president when I was still chairman and CEO. In the spring of 1986 he became CEO. Then in the summer of 1986 I left. In the 1984 to 1986 period he assumed increasing amounts of management responsibility. Divisionalization was his initiative. That was the one way you got some leverage and some scale. It was by taking the same basic product and technology and serving different markets in the different divisions. I don't think that really worked out. There was an Engineering & Scientific Products Division that just did not work. The products were not right and weren't delivered. That divisionalization scheme went away. I don't even know how they have it divisionalized today. It's been through multiple reorganizations. This is a classic management problem of how to deal with complexity, and one way you do it is to divisionalize. I have to say that I'm just not very good at understanding how one ought to structure a large organization. I'm a start-up kind of person.

Goldstein: Was it tough for you to relinquish control at various stages?

Kapor: Let me tell you what it was like. As I said, the whole thing was like being in a war zone the whole time. It was enormously stressful, very exciting, and with a great deal of gratification and success. I was really afraid of screwing up a lot of the time. As you can gather from what I've been saying, I was also extraordinarily ambivalent about running the place. Ultimately I decided I would be happier if I wasn't. This is not a decision I have regretted at all. I tend to be a perfectionist, and I can be very control-oriented. If things are not the way I think things should be, then I get very unhappy. I became very unhappy because I was unable and unwilling to run things and unhappy with the results of not running them. So it was a great personal dilemma.

Goldstein: You are providing two contexts for 1986. One is the maturation of the business in the industry, a different place in the life cycle. The other is your personal position and control in Lotus. Did those both contribute to your decision to leave?

Kapor: Yes. I felt that there was just too much responsibility. I couldn't sleep at night. I think I felt overly responsible for the whole thing. That certainly didn't help. It wasn't my ambition to run a big company. I wanted to do this great product and make a big business out of it. But I didn't find the positive parts of running this big show to be very gratifying. I mostly felt that there were a lot of people who wanted a lot of things. This was whether I was a CEO or not. I was just being the chairman and the figure head and so on. They wanted things all the time, and it was just a pain. I like to be left alone to do my own thing. But instead, I was a prisoner of the spreadsheet. It just got to be this big monster. I could see that Lotus was going to be the company that was going to be dominated by the spreadsheet for many years to come. I like spreadsheets, but not to the exclusion of everything else. It's like being told you can't eat anything but Chinese food for the rest of your life. Who wants to do that?

Goldstein: Did you try to diversify Lotus's product line?

Kapor: We tried hard, but in a strategically stupid way. One thing that Lotus did that other companies did not do is systematically experiment in new types of products and bringing them to market. The only one that really looks like it's going to work is Notes, which is going to become the cornerstone of the business. But I did Agenda, which got great critical reviews and has fanatically appreciative users, but which was not a commercial success. Then there were all these other products. There was stuff we acquired from Software Arts. Metro was a pop-up TSR. And then Lotus Express; we had an early e-mail package that was a smart front-end for MCI mail. We had Symphony, we had Jazz, and we had lots of things that never made it to market. We had product after product after product. In fact, we were known for this. People would say, "Great ideas for individual products; no coherence." We did Manuscript, which was an outline-oriented word processor. If we had been simpler and less ambitious in the product spacing—let's just try to do a decent word processor, let's not try to reinvent the category here—and if we'd started on that early and just kept improving it slowly we would have done much better as a business. That was the applications strategy at Microsoft.

As it is, Lotus got into the word processing business by acquiring Samna. Omni is very good and gets the best reviews, but it's competing against the entrenched competition of WordPerfect and Microsoft Word. It was strategically very weak, but creatively very inventive.

Goldstein: Did you introduce all these products and try to diversify the way you did because of your instinct?

Kapor: It was because of my instinct and my desire to put out good products.

Goldstein: Or, did you conceive of it as a wise business strategy?

Kapor: No! The former. I didn't have a clue, really.

Goldstein: It was just this compulsion to make good stuff?

Kapor: Right. This is not atypical. This happens.

Goldstein: I read that you like the Electronic Frontier Foundation because it is like the fun days of Lotus.

Kapor: Right.

Goldstein: Can you elaborate on the differences between the fun days and the days that were less fun?

Kapor: It is just what we have been talking about. When it's small and you are running it you have this nice team feeling and you can know everything that's going on. You really feel like you are making a difference. Things happen quickly. There is a lot of novelty in it. When things get to be bigger every move you want to make has many consequences, both internally and externally, that slow down the pace of things. There are a lot of things that people tell you you cannot do or should not do because it's going to have some negative effect on the business. Part of the problem was that I think I listened too much to people who didn't see the world the same way that I did. I didn't trust my own instincts.

Goldstein: Would you characterize On Technology as a start-up?

Kapor: Yes, it was.

Goldstein: How did the experience in starting it up compare with Lotus?

Kapor: It was a struggle. I made a series of fairly classic mistakes when entrepreneurs start second companies. I was overly ambitious and did not have a sense of appropriate limits. Even though I was aware that those were risks, and even though I took those risks seriously, I still made the mistakes. You try to do too much. Once again, I wasn't willing to be pragmatic in the business sense. I had this vision of product purity in mind. We raised a lot of money on high valuation, hired a lot of smart people, and then.... I'm also a pragmatist. That's the problem. I can only fool myself for so long. Some people can fool themselves forever. I realized it was not going to work anymore. I didn't have the fire in the belly to do the system software, the next great operating system. It wasn't my real expertise, and I did not want to compete with Microsoft. We tried changing the strategy to an applications strategy, but I had to fire a lot of people, and it was just a mess. I really lost my will. I shouldn't have started another company. I had done everything that I was going to do. A lot of it has been trying to overcome the bad start.

Goldstein: Could you explain what you mean when you say the classic management problems of starting a second company?

Kapor: Many people who were successful with a first company and start a new company want to prove to themselves and to the world that the first company wasn't a fluke. Since they were the big success the first time, they feel they have to be an even bigger success the second time to prove it was not a fluke. But that interferes with letting things take an organic course. Starting something to prove a point is not a good strategy. Now look at Next. Steve Jobs started Next. Black is the color of revenge. He was thrown out of Apple and then went and did things that were just self-defeating. That is really what I mean.

Goldstein: You refused to learn the lesson of big companies that you had left.

Kapor: Right. He also did something that was overly ambitious and not market-focused or pragmatic. It didn't work.

Goldstein: What kind of continuing education did you provide at Lotus? Was that an issue for people who worked there? For the employees?

Kapor: We had a tuition reimbursement plan that was pretty generous. People picked their own courses if they were job-related. Eventually the company developed a bunch of seminars and other things. But believe me, we were just scrambling to keep up with things.

Goldstein: What qualities did you think were important in the employees? As you expand, you have to bring on a lot of people.

Kapor: While I was there, I was very concerned with the quality of work life, seeing that it was a good place to work and that people were treated fairly. We made a lot of investment in the human resources function and in programs to support that. What I found was that at our rate of growth, we attracted a lot of opportunistic people who were there who really were not the best for the company. We got a culture that was a culture of entitlement to an extent.

Goldstein: What qualities were you looking for?

Kapor: What we really needed were people who were smart, fast, and humane.

Goldstein: How did you try to use ties to either academia or the government? Did you make alliances?

Kapor: We really did not do much of anything with government. There was nothing formal with academia either. We were just being entrepreneurial.

Goldstein: What were the decisions about the products that were made? How did you set yourself to keeping an ear to the ground?

Kapor: In the earliest days my intuition and my personal knowledge was really what we used, and it worked well. We tried to build in a lot of customer contact to understand what customers wanted both informally and formally. We made site visits and had focus groups. I think that is the right thing to do. Unfortunately, a lot of what we heard was not what we wanted to hear, and this created various conflicts. We were not strictly dedicated to doing what was going to make the customers happy. That would have been a much better business strategy. We had these conflicts between a vision of doing things with new kinds of products versus what customers wanted. That was never successfully resolved.

Goldstein: Do you have any comments just on general management lessons?

Kapor: Managing small scale and managing large scale are two very different things. When you can know everybody, whether that is five people or a hundred people, you can lead by personal influence in a way that is very difficult to do when things are larger. I admire what Bill Gates has done at Microsoft by staying very closely aligned with the key developers—probably a couple hundred of them. He has enormous influence throughout the company, determining its product direction and its strategic direction. I think there are huge portions of the Microsoft empire which he just leaves completely untouched, that are managed through other people. I have a different product philosophy and a different set of values. You have to admire the kind of job that he has done in maintaining this effective span of control of this enormous empire. This could be focused in on empowering the developers and staying closely aligned with them, and driving the business well around that. That has worked really well. There's a real lesson in that. It's the only way I know of scaling things.

Goldstein: What was your attitude about research and development at Lotus?

Kapor: I loved R&D. I would have been happy just to do R&D. A lot of products we put out were really R&D. That is where the interest in new ideas is, and that is where the smart people go.

Goldstein: Did you try to integrate or incorporate technologies from other areas into your R&D? I'm thinking particularly of artificial intelligence.

Kapor: Yes we did. It helped Agenda to an extent. We looked at a lot of that. Some of it went into Improv. We tried to hire people with a bit of a research background and who were current with that kind of stuff. When I was there, I always tried to get developers who were interested in producing products, not people that were interested in writing papers. We were not big enough to afford that.

Goldstein: I only have a weak sense of what R&D is like in a software company.

Kapor: Apple and Microsoft and their advanced technology groups will play with things like speech recognition, and they will try to get a project to demonstrate that it's feasible to do a certain type of thing, such as having useful speech recognition for some class of applications. Then they will do a project that is a prototype to demonstrate and show them. So that is less research and more advanced development, where you take some technology, whether it's AI or speech recognition or some computer graphics technique, and you try to take it the next stage, toward pragmatism.

Goldstein: Does that lead inevitably to a vertical integration? Or do you have to get involved with hardware?

Kapor: Not necessarily. You might, but you might not. Often, especially with the improvements in microprocessor performance, you can run many things with off-the-shelf hardware. It's just that particular algorithms and data structures that are used in AI are a thing unto themselves. If you do not have people who understand that world and what it's trying to do and how those techniques work, you cannot possibly use that stuff.

Goldstein: Would you prefer to shy away from that kind of vertical integration? Or does that seem interesting?

Kapor: This is very interesting.

Goldstein: Did Lotus do any of it?

Kapor: You mean with hardware?

Goldstein: Yes.

Kapor: Not really.

Chapter 11

Arthur P. Stern

About Magnavox

Magnavox began in Napa, California in 1911 as Commercial Wireless and Development Company (CWDC), a research and manufacturing company in the field of wireless telegraphy and telephonic communication, with $2500 capital. The company's first great success came in 1915 with the development of the electrodynamic loudspeaker. A year later CWDC invented the first electric phonograph. In 1917 CWDC merged with Sonora Phonograph Distributing Company to become the Magnavox Company. During World War I Magnavox entered the defense market with antinoise communication systems for air-to-air and air-to-ground communications for open-cockpit planes. Magnavox was a primary producer of radio and telephone communications systems during World War II. In 1951 they formed a separate Government and Industrial Division to indicate their long-term commitment to serve the Department of Defense.

Concurrently with their advances in military projects, Magnavox developed numerous innovations in the field of consumer electronics. They developed a radio/phonograph unit, the single dial radio, and hi-fi radio—all prior to 1940—earning themselves great recognition in the commercial radio industry. Their consumer electronics work soon extended into television. By 1969 Magnavox had developed the first automatic color television.

North American Philips Corporation purchased Magnavox in 1974 and then divided the company into two divisions: Magnavox Government and Industrial Electronics Company (MAGIEC), and Magnavox Consumer Electronics. North American Philips consolidated Magnavox Consumer Electronics division with its Sylvania and Philco companies in the early 1980s. In 1991 MAGIEC was renamed Magnavox Electronic Systems Company.

Today, Magnavox employs more than six thousand employees at corporate facilities in seven American cities and maintains its headquarters in Fort Wayne, Indiana. The company continues to be a leading supplier of military and commercial products in communications, antisubmarine warfare, infrared systems, satellite navigation and communication, electronic warfare, and tactical information systems. Magnavox is the world's largest producer of handheld radios and is the leading supplier of underwater detection devices. Many Magnavox product lines are currently in use by the Federal Aviation Administration, commercial air carriers, and the National Aeronautics and Space Administration.

Arthur P. Stern

Place: Edison, New Jersey

Date: July 13, 1993

Aspray: Why don't we start by having you tell me something about your background: where you were born and what your early education was like?

Stern: I was born in Budapest, Hungary, in 1925, to a religious orthodox Jewish family. I went to elementary school, and then to middle school, both orthodox Jewish schools. Then I went to yeshivas, that is, prerabbinic schools, up to the age of seventeen. While I was doing this, I was also pursuing my secular studies. I finally joined the Gymnasium of the Budapest Jewish Community, where I graduated in 1943 with what was called a Maturity Certificate. I've just come back from Hungary from the fiftieth anniversary of that. I studied law in Budapest from 1943 to 1944. In 1944 our Holocaust started and we were deported to the concentration camp Bergen Belsen in Germany. From Germany, toward the end of the war, I got to Switzerland. It was there that I decided to switch to engineering. I studied at the University of Lausanne and then at the Swiss Federal Institute of Technology in Zurich, where I obtained my engineering degree in 1948. A Swiss "professional" engineering degree (Dipl. Ing.) is somewhat similar to a master's degree in this country.

Aspray: What was the field?

Stern: Electronics. I think you can call it electronics, but electronics was very different at that time. It was electrical engineering with some specialization in high-frequency technology. I then worked for an industrial firm in Switzerland. I taught at the Institute of Technology in Zurich for about a year and a half before I came to the United States in 1951 and joined General Electric's Electronics Laboratory in Syracuse, New York. I had the exhilarating experience of participating in the development of the country's color TV system from 1951 to 1952. It was a wonderful opportunity to make some contributions.

In late 1952 I felt that transistor technology was far enough advanced to develop a transistor radio receiver. I was appointed project engineer on that job by the management of GE's Electronics Laboratory. In early 1954 we were the first ones to demonstrate and publish a comprehensive paper on transistor radio receivers. Shortly thereafter I became manager of the GE Electronics Laboratory's Advanced Circuits group. It was my first foray into management. I was responsible for the development of all kinds of solid-state techniques, systems and subsystems in that capacity.

In 1957 I became manager of the Electronic Devices and Applications Laboratory, which gave me responsibility over things like lasers, masers and various solid-state devices, semiconductors, magnetics, dielectrics, et cetera. I stayed in that position until 1961, when I left GE to serve as director of engineering of Martin-Marietta's Electronics Division in Baltimore.

Aspray: Why did you make that move?

Stern: I got fairly high up in the research and development hierarchy of General Electric. I was sort of boxed in. My salary level was too high to switch over into engineering without taking a loss in grade. Then Martin-Marietta came along and offered me the management and development of a large engineering organization, just on their faith that I would be able to do it even though my entire background was in R&D. So I took it, and that was an interesting thing. I stayed with them from 1961 to 1966. It was a very good change. The Martin-Marietta Electronics Division merged in 1964 with TRW's Computer Division. These two together formed a jointly owned subsidiary called Bunker-Ramo Corporation. I became director of engineering and then director of operations of Bunker-Ramo. The latter position included responsibility for everything except human resources and marketing. I was number two there like I had been at GE since 1957.

Then Magnavox came along in late 1965. They had the Magnavox Research Laboratories on the West Coast, and that was an opportunity for becoming number one. I switched in early 1966 from the Martin-Marietta empire to Magnavox, and I stayed with them until my retirement in January of 1991. I spent about twenty-five years there.

During my association with GE, after I switched from color television to solid state, I had an opportunity to be a member of the group of people, centered mostly on IEEE, that promoted and developed solid-state technology throughout the country. There were maybe thirty or forty people throughout the United States who were leading in that area, and that was a tremendous experience. Later at Martin-Marietta I was involved more in the development of substantial military systems, antisubmarine warfare, missile guidance, and command-control systems. At Magnavox I had an opportunity to start new things and see some of them through. In particular, we established spread spectrum communications in the world. It was quite difficult because you needed a lot of components and integrated electronics wasn't around yet at a sufficiently high level of integration. It was tricky to keep these very complicated systems and subsystems working for more than a few hours at a time without failing. But then, of course, as solid state came along, these things became very practical. Today everybody uses spread spectrum technology.

The next interesting challenge was satellite navigation, where we at Magnavox became world leaders. We brought satellite navigation to commercial shipping and a variety of other commercial applications such as high-precision position location for surveying. Then came the next generation of satellite navigation, which was the GPS system. GPS stands for global positioning system. We started it conceptually under a subcontract with Hughes Aircraft. There we built the first practical demonstration of system feasibility. We won several competitive contracts starting in 1968, only to lose the big production contract to Rockwell in 1985, seventeen years after we started. But it was an exciting experience. We did a lot of new things. I think I was very lucky to be associated with people who enabled me to make a small impact on the lives of many people. I retired in early 1991, and since that time I've been consulting and participating in a variety of volunteer and committee activities.

Aspray: Can you tell me about the nature of Magnavox's business? How was it structured? In what areas did it start?

Stern: Magnavox was established, if I'm not mistaken, in 1908. Within the past decade we had our seventy-fifth anniversary. It was named Magnavox, which comes from Latin and literally means "magna vox" or "big voice." They were in the loudspeaker business. Then from the loudspeaker business they went into a variety of other areas. During the Depression years it was difficult to survive. They almost declared bankruptcy in the late 1930s. Then World War II requirements for electronics rescued them. They became a substantial military producer. After the war they went into television and later into color television, while maintaining their defense activities.

By the time I joined in early 1966, the company was in home electronics, including color TV, hi-fi consoles, etc., a fair amount of military and industrial electronics, and some non-electric activities such as furniture. Furniture was very important for electronic consumer products at that time. In those days the hi-fi set would be a large nice-looking furniture console with some electronics in it.

Aspray: Yes. My parents had such a Magnavox product in their home.

Stern: Then things were difficult in the early 1970s because of Japanese competition. The Japanese had a better understanding of what the consumer market needed and made many technological innovations. They implemented solid state in TV at a time when Magnavox still used tubes. They came up with modular consumer electronic systems, while Magnavox still tried to sell big pieces of furniture. Things got tough. The company was very profitable at the time I joined it and for a few years thereafter, but then it started losing money.

Philips, the Dutch electronics giant, via its North American Philips arm, made a tender offer for Magnavox and in 1974 acquired Magnavox *in toto*. Philips did this in order to acquire Magnavox's consumer electronics business, and at the time they were hardly aware of the Magnavox defense and industrial electronics businesses. We told them after they acquired us. But they were very good to Magnavox for many years. They gave us a tremendous amount of freedom. They chopped up Magnavox into a variety of companies. One of them, the Magnavox Government and Industrial Electronics Company (MAGIEC), dealt with the defense and industrial business. I was first a vice president and then a senior vice president of that company. My organization grew from a laboratory to a pretty sizeable product organization between 1966 (when I joined) and 1980 (when it was established as a separate company). I then became president of what became known as Magnavox Advanced Products and Systems Company (MAPSC) and stayed there until my retirement.

Aspray: Why did Philips decide to break it up into separate companies?

Stern: Philips management principles at the time called for very limited interference into the businesses that they owned. They were delegating as much of the responsibility as possible to the local management of the business. Philips management at the central level recognized that they were not experts in many industries in which they were engaged. In the United States, Philips was a real conglomerate. (Later, in 1989 to 1990 when Philips got into financial trouble, they sold most of their noncore, peripheral businesses.) So it was in line with the company's thinking to delegate a lot of responsibility and in turn to encourage businesses that could stand on their own. They would give the management of a business authority to do whatever they could do and then hold them accountable for the results. After acquiring Magnavox, itself somewhat of a conglomerate, Philips divided Magnavox into its component business elements and gave each its autonomy.

Aspray: What sorts of things were centralized within the Magnavox group or within Philips itself? Research? Capital? Marketing?

Stern: If you consider the period starting with 1974, at the time of Philips' acquisition of Magnavox, and ending in the late 1980s, Philips really did not involve itself in any significant business decision, except when we needed substantial capital. We went to them, and they either made the capital available or not. They generally did. As I recall, they hardly ever didn't. If we wanted to do something big, like building a new facility, they exercised the authority to decide whether to own or lease the facility. In other words, they assumed a more active role whenever substantial capital amounts were involved. Otherwise, Philips did not interfere. As a matter of fact they couldn't because, as a predominantly defense business, we were under the special supervision of the U.S. Department of Defense (DOD). Philips was not even supposed to know (and they didn't know) about the classified things we were doing. By agreement with the DOD, they were not allowed to interfere in any way. They behaved like a bank. We would simply give them a major presentation at the end of each year on what we intended to do next year, mostly in financial terms. We would give them a little information about technology, whatever was not classified. We also gave a mid-year presentation which was shorter. That was about it. We were governed by our management and by our board of trustees, which was appointed jointly by Philips and the DOD.

The central Magnavox management was more active in managing things. There were several companies and organizations that reported to the central Magnavox management, including mine. The central management allocated part of the research and development funds for us. They also approved or disapproved major bids—those in excess of $10 million. Otherwise, during the last ten years I ran my company pretty independently.

Aspray: I don't know much about the defense business. What are the kinds of things that make a defense company successful?

Stern: There was considerable difference between the two major components of Magnavox: the so-called Magnavox Electronic Systems Company (MESC) headquartered at Fort Wayne, Indiana, and my company, MAPSC, headquartered in Torrance, California. MESC was 95 percent defense while my company was about 65 percent defense and 35 percent industrial. I really had very little to do with what was going on elsewhere. In my company the backbone of the defense business was spread spectrum systems and subsystems. We were responsible for applied research, technique development, product development, and ultimately the production of some major military communications systems. In particular we developed, designed, and produced the core of the country's successive strategic communications systems: the AN/URC-55, AN/URC-61, and the AN/USC-28. The systems we designed had all kinds of ramifications. For example, when the president of the United States—don't forget that those were Cold War days—left the country to meet with Gorbachev in Iceland, he required instant access to the country's strategic assets, including the nuclear capabilities. So Air Force One carried one of our systems, which made it possible for him to have access at all times to the levels of military management with which he had to be in contact.

The strategic communications network was rather extensive. Around the world there were many terminals. It was expensive, very important stuff. This system used protected communications. Protected communications means that you can communicate with your friends without the enemy being able to interfere or listen in on what you are talking about. The enemy or potential enemy would not even know that you were communicating. This is the kind of system that we developed and manufactured.

Aspray: What did the business of that division amount to?

Stern: That division was doing somewhere on the order of $80 million a year. Not very large but nevertheless a sizeable business. It was quite profitable and always on the edge of the technology because improvements had to be generated all the time. Of course the military budget in the 1970s and 1980s was sufficiently large that these almost continuous improvements could be sustained financially. Thus multiple generations of these systems came about, and the equipments that we built in the late 1980s were not comparable to the equipments we built in the mid-1960s—although the fundamental principles remained the same.

Aspray: How does a division like that plan for future developments in a product area?

Stern: Obviously, any organization engaged in that type of activity must constantly be aware of technological trends. It is impossible to be successful in a business of this type without knowing exactly what's going on and without being able to do something about it. You are engaged in continuous technology monitoring and technology forecasting in a very practical fashion. You know that the customer expects to use our equipment for somewhere between five and ten, maybe even more, years. You know that the customer would not view your company favorably if you pursue major modifications every six months. But you also know that the customer will consider you unreasonable and not really on the ball if you didn't come up with potential improvements within the next year or two that will be ready in three to four years.

So you have activity all the time. You are constantly evaluating new technology. You are forecasting which parts of the new technology will last, which make sense, and which don't make sense. Sometimes you make the wrong bet. Sometimes you do not perceive that a new technology really impinges upon you, and then you are late. In other cases you bet on some new technology that doesn't make it. We had our share of both. It's a continuous process; it's a way of life. The function of top management is to allocate resources in such a way that your access to new technology is assured, and that you have the right people, who in the majority of cases make the right decisions and bet on valid kinds of technology. This is not necessarily a very efficient process. It takes people of different abilities and you have to be careful of the type of abilities that you are accumulating. You also have to be very careful of the type of people because some people are tremendously productive and others are not. They may have gotten the same degrees in the same year at the same institution, and the one who is not productive may have had a much higher grade point average than the other who's productive.

Aspray: Is there a special set of techniques that are used for doing your forecasts?

Stern: I don't know how special these techniques are. People who are in top management in a highly technological organization, while not necessarily technically creative, must be on the ball and must know what is going on. I always considered it one of my responsibilities to know what's going on technologically.

Aspray: Did most of your senior managers have engineering backgrounds at some point in their career?

Stern: No, I wouldn't say that. My chief financial officer did not. But even in that kind of a position it is good to have some technical understanding, and the bright person is going to pick it up very fast. You have an important combination of people who have the technical savvy, the manufacturing savvy, and the marketing savvy. In order to succeed, they'd better know everything. There will be a redundancy of knowledge and a redundancy of objective which they can work out among themselves, obviously with my participation. Then when the chips are down, the person responsible for the whole thing—in our case, it was me—will make some good decisions and some bad ones. I made both. I made some outstanding decisions and I made some outstandingly lousy decisions.

Aspray: What about technical monitoring? How do you do that?

Stern: We had a number of excellent technical people on our staff. We conducted design reviews for every activity on a regular basis. That must be part of life. You cannot let technical groups of people go off on their own. Not because they are not bright. The reason actually is that they may become too bright. You must have a number of crucial things in your organization if you want to be successful. One of them is standardization, and that means that you ought to tell the bright engineer, "I know you want to use this component, or this approach, or this tool. And I know that it appears to you as superior. Who knows? Maybe it actually is superior. But we can have only so many approaches inside any organization. Otherwise, we are fractionated. So, let me tell you something: We are not going to use that." That's part of life. Incidentally, that results in resignations and in upset and angry people. That's part of dealing with technical people. It's exciting, it's productive in a practical sense, and it's very rewarding. Sometimes it's absolutely frustrating. The young engineer says, "But I know that this is better than what we have used." And you have to be firm, "Leave it that way. You may very well be right. As a matter of fact, I know it. But we can only have so many approaches. And this is one that I'm not going to use. I'm going to wait for the next generation and skip that one. I'd rather skip two stairs than one." Many young people will not accept this.

Aspray: What about monitoring what's going on in the government laboratories, the academic sector, and your competitors? How do you do that?

Stern: If you are set up reasonably well, that's not a problem. Are we still talking about defense electronics?

Aspray: Yes.

Stern: As I mentioned, that was two-thirds, and in fact, the less interesting two-thirds, of my organization. I was spending most of my time on the one-third that was commercial. But monitoring is not much of a problem. If you are set up properly, your marketing people will be dealing with a variety of government customers and will learn from them what's going on and what's being planned on the system and subsystem level. On the techniques and technology level, your engineering people will follow up on their own or because you insist on it, and they will participate in relevant meetings of technical societies such as IEEE, NSIA, et cetera. But I should say that the IEEE is no longer the home for engineers that it was in my time. The members will follow what's going on technologically. But a lot of them just read what the magazines have, and it's often outdated by the time it's published. The leading people, however, will go to these meetings, and they will meet people informally.

They will know their peers in other organizations. And in no time you will know what's really going on, well before the publication of it.

I didn't find that with an organization of reasonable size and reasonable emphasis on gathering technical intelligence that it is terribly difficult to know what is going on. You do this legally. It's all clean and there is nothing improper about it. To me the problem never was knowing what was going on. You knew it practically as soon as it was thought of—with some exceptions, of course. I don't want to exaggerate here. There were very few things that were successfully kept quiet by some companies. Usually they don't succeed because people are migrating, and this is a free society. More important, people who are plugged in talk with each other. The problem is how to have a modicum of control and what to do about it. Marx said—he is not frequently quoted today—well over a century ago that German philosophers have evaluated and talked about the world for years. The issue is not what we talk about. The issue is what we do about it. That's the situation with technical management: What do you do? Do you sit at these meetings and pick up information or have your people come in and tell you, "I picked up this, I picked up that." There comes a point when you have to say, "Now we *do*." And that's difficult.

I recall one of the more important success and failure stories of my career. In 1969, when our Fort Wayne sister division failed to exploit satellite navigation profitability and when the company's top management was about to terminate the satellite navigation activity, they decided to offer it to my division if we wanted it. This was before the acquisition by Philips. It meant moving it from Indiana to California. We decided at that time that this was very exciting. So starting in July 1969 we were engaged in satellite navigation with the objective to make it a commercial product. Just to give you a little background, satellite navigation was really established by the U.S. Government, with the intent to equip the country's Polaris submarines and their ballistic missiles with precise position-location and navigation systems. When you shoot something off and you want it to go exactly where you want it to go, you've got to know where you are with high precision. This is not only good advice for life in general but it is absolutely mandatory for missiles. That which you deliver will arrive with lower precision than the precision of your location. So if you've misjudged your own location by two miles, the missile will not fall on the target but two miles off—at least. Then there are additional errors which accumulate placing the missile five miles or ten miles away. So precise position location from the point of view of moving ballistic missiles was essential. Johns Hopkins University's Applied Physics Laboratory developed a very fine position-location and navigation system using satellites. They were called Transit satellites or the NNSS—Navy Navigation Satellite System.

This was a tremendous opportunity. Vice President Humphrey announced in 1968 that we will make a large part of this technology developed for Polaris submarines available to the world at large for commercial applications so the world can benefit from these improvements in satellite navigation. So we said, fine. We went for that. ITT made the same decision at the time. This became—between 1969 and 1973—an area of intense competition between ITT and us. We finally beat them out, and then there was just us. We made a lot of mistakes in doing this. We did the job, but we did not become a big commercial success. The market wasn't ready. It took a lot of learning.

I remember my first trip to Norway in 1970. At that time Norway was a major power in commercial shipping. We met with one shipping company after another to learn what they really wanted in precision navigation. We learned after a few fascinating meetings that in each shipping company there were two powers: the director or vice president of Operations and the director or vice president of Engineering. The guy in Engineering always wanted the best. He wanted complete control. The vice president of operations said, "No. The eighty-three captains that I have around the world on eighty-three ships don't want us to know where they are and what they are doing all the time." The director of Engineering would say, "I want a permanent record. I want the printout of where this ship was all the time." The director of operations would say, "Look, you're crazy! You don't understand! The captains will sabotage the system. We don't want this thing." We learned that our systems had to be satisfactory not only from a shipping company's technical viewpoint, but also from its internal politics perspective.

We also learned that military reliability was not good enough for the commercial shipping people. If you're going to use our equipment on a ship, and it fails within a year, they'll send it back and never order any more. Formal acceptance doesn't really mean a lot. If they accepted it and it fails, they won't reorder. Whereas the military at that time, if they accepted it and then it failed, they would pay you well to repair it. So, there is a big difference.

For two or three years we concentrated on the competition with ITT and learning about the shipping industry. After we won, we still didn't have products that were sufficiently popular. We had an engineer in our organization then who was really bright. He said, "You know, we've got to come up with a system using microprocessors." Microprocessors were quite new then. We got together, debated it, and agreed to do that. So in mid-1976, we came out with an instant success! It was the sort of thing that is almost a once-in-a-lifetime event: You develop the thing, you sell a few of them, and you get wonderful feedback. Suddenly you don't know how you're going to fill the orders. That's what happened to us. We were in a dream in the second half of 1976. We tried to increase our capacity. We were somewhat lucky because I decided that in the original run we would produce 250 instead of the usual fifty.

By the end of 1976, we knew the product was a tremendous success. I got the guys together, and I said, "You know, this is great! And it's going to last for a while, but everybody knows about it. Somebody is going to come up with something better within the next two years. Therefore we're going to design something new and do something much better." We established the priorities on what this "something better" needed to be and development started.

Let me make a long story very short. I became involved in many other activities. I did not check on the progress. I assumed that the written directive which defined in fair detail what should be done would be followed. I found a couple of years later, when we were supposed to have the product, that we were far from being through, and furthermore the directive was ignored. A number of expensive technical "innovations" were introduced that shouldn't have been there. These were bright things—new, daring, super high-tech—but too expensive. We had to regroup. Instead of two years, it took us four years, including a major redirection after two years. It was a catastrophe. We did finally recover from this, but we had several down years in the meantime, and we had a loss of market share which should have never been.

Why do I relate this? It was one of the really bad things in my career. You cannot walk away from a project. You cannot say, "Well, I've given instructions, and everybody agreed, and it's going on." Unless you constantly monitor, unless you constantly emphasize, unless you issue minor redirections every two or three months as they may become necessary, you are bound to have problems. If you have looked the other way for a year the redirections will become major and you will have lost time. That's the way life is.

Aspray: In regard to the defense business, I have this impression of business lock-in. Once you get in with a system, you continue to modify it and improve it. You have a great advantage over your competitors in continuing business with the government. Can you talk about that kind of market and how you protected it?

Stern: Well, to a large extent it is true. At least it was true during most of my career. Toward the end of my career it changed somewhat. Each business has its traditions and its ethics which are not necessarily the ethics of others and of other businesses, and which sometimes are not even rational or in conformance with general ethics. But a business has its practices, which then become its ethics unless something happens—which actually did occur in the 1980s when the Department of Defense suddenly declared certain practices of the defense industry illegal by enforcing laws which were previously ignored. Some people went to jail or were fined. But that's also a question of monitoring. When you have laws, you've got to monitor your organization for compliance with the laws. The fact is, for a long time during the first thirty-five years of my career in the defense industry, from 1951 when I joined GE until the mid-1980s, the key to success was to win the R&D phase of a new system. You did that by underbidding it. You could overrun it if it was a cost-plus situation, but if you overdid it, that made the customer very angry. So you frequently decided to swallow much of the cost overrun even though the contract was cost-plus because you didn't want to make the customer angry. Typically, you lost a lot because either it was a fixed price and you had to lose, or the customer relations were such that it was better for you to lose than to compel the customer to come up with the last penny he had, thereby earning his undying enmity. But once you were through with that, nobody had a chance of taking the production away from you. And on the production you did well. From the initial production you recovered that which you lost in R&D, and then as things went on for years, you made a lot of money.

This approach was supported by a number of things. It was supported by the procurement methods that were used. What do I mean by that? First of all, a low R&D bid generally got the award, even though everybody, including the evaluating government officer or team, knew or should have known that you would be losing money on it. Or, there would be an overrun and they would have to come up with more money. They had, in principle, the right to say, "Oh, you are underbidding. You are overly optimistic. Therefore, we are going to give the job to the other guy." It happened on rare occasions. It was a great exception. Suppose one was coming in with a fixed-price bid of a million dollars; they would suspect it's going to be two or three million, but they aren't going to do anything about it. So that was one procurement practice that contributed to this kind of thing.

The second one was tricky from a legal viewpoint. When you bid the production phase, which was for a fixed price in most cases, you had to face price negotiations. Price negotiations are a very sensitive thing because you must comply with truth in negotiations requirements. That means if in the system you are going to use six widgets and you have the vendor bids for these at, say, one dollar each, you put them into my bid at a dollar each. If this is then agreed on in the negotiations, there is a limited profit on that production element, say a profit of 10 percent. So you get one dollar and ten cents for every dollar cost. But you may suspect that, when the chips are down, and particularly in the second and third phases of production, you're going to buy these things for sixty cents each. If you have evidence that it will be sixty cents and not one dollar, you must reveal it to the government during the negotiations. But if you have no evidence, you don't have to reveal hunches. So you can make a tremendous amount of money because you have this production for years. You are buying these things at the beginning for a dollar a piece. Then you buy them for sixty cents. Sometimes you may buy them for twenty cents. The government would still pay one dollar and ten cents. And with that, you made up the losses that you had in development—and much more. It was a very good business. This is of course a highly simplified example. Reality was more complicated but, in essence, the same and very profitable.

Aspray: But that means that your parent company has to have fairly deep pockets for a while because the payoffs are pretty long term.

Stern: That's right. That's the way it worked. Much money had to be put into certain things until the money came back. But then it came back in large amounts. Obviously a small company couldn't afford to do that, and obviously this all tilted things toward the big companies. These were the practices, and there were frequent arguments with the government about truth in negotiations. A very difficult situation which we had was the following: Out went solicitations for this cup which we needed. We got an offer from one vendor for a dollar each. This was incorporated into our bid. The bid went out on July 1. The government found out that one of our engineers, without reporting it to management, had already discussed on June 25 with another vendor the possibility of changing this cup slightly so that it would cost fifty cents. These were often quite honest mistakes. But communications inside the company were not that good. So there were a lot of misunderstandings. All this then changed in the mid-1980s when the government became very touchy about things, particularly as available defense funds decreased. Today things are much more complicated because the government is more precise in its evaluations. Life has become more difficult and more bureaucratic.

Aspray: In the defense business, how would you expand your business? Did you try to build up your market share? Did you just assume that you'd be making more money on these products that you already had?

Stern: It depended on the kind of company that you were. There were always companies who were primarily production companies who were built on taking the production away from the original developer. For example, Magnavox developed the VRC-12 army field communication system and produced it for I don't know how many years. Then somebody came on the seventh or eighth production run and underbid us. They took the business away from us. These were usually comparatively low technology and therefore low overhead companies who were good in manufacturing techniques and were able to manufacture inexpensively. Full-range companies started with applied research and stayed in the business through initial production and services or even longer. Once something was somebody else's, we generally forgot about it. We didn't go after that because it was a bad deal. I would usually not go after something that Hughes Aircraft had already had. But that's why you have marketing people and you have technical people. You could get into it in two ways: One is that you internally initiate new functions based on the needs of the government as you perceive them. If you proposed things that were satisfying a need and ultimately satisfying a formal requirement, and if your proposal was a reasonable, near optimum balance type proposal, then while others will have an opportunity to bid, you have a good chance of getting that contract. From then on you're in. In other cases, the need was identified by the government rather than by you. You learned about it relatively early because you have marketing people with contacts in the most important government agencies. Then you started working toward satisfying the need. You worked with the government in developing technical methods by which the need can be satisfied. But in both cases, whether you originated the lead or the government initiated the lead, you were knowledgeable of the need and proposed something new. So you expanded your business by getting into something new.

One example of the exciting things that we got into was just at the time when we lost the big GPS competition to Rockwell. We bid on a project for which the government had a need, which we perceived very early. It was for a field facsimile system. Many Japanese companies make facsimile systems, but I'm talking about a military field facsimile system. That's a facsimile system that you can drop from a helicopter, and it will still continue to work. If you work in a physical environment which is very hard, you can't buy it for five hundred dollars. It will cost many thousands of dollars. We proposed this, and it happened to be in 1985, just at the time we lost the GPS. The government decided in our favor in the competition. That became a major multi-hundred-million-dollar business. We had developed the first version military fax in 1972 or so. Then we modified it and modified it again. We sold a few to some odd government agencies. By odd I mean not the ones that buy a lot. Then we sold it overseas. I remember the Israeli Army bought some. Then came the serious interest of the U.S. government, so we redesigned the whole thing. That which we produced had nothing to do with the various models that we had already sold. We finally sold to the U.S. Army, and created a lot of new technology at the right time. Timing is very important.

Aspray: You said earlier that two-thirds of your responsibility was in the defense business and one-third of it was in an industrial business. Do you want to talk more about the industrial side?

Stern: Important parts of the industrial group were the satellite navigation business and communications for ships, primarily. But then we went after a variety of other types of vehicles and also for precision survey instrumentation. It was very inspiring because it was really new. At some point I got sick of the military marketing process that I described earlier. The commercial business was pure competition and more exciting. We did the right thing, and we knew what the results were. In some cases we lost. In many other interesting instances, we had tremendous success. For over a decade starting about 1976 we had the majority of the world market in both satellite navigation and in precision position location.

Aspray: Was the product life expected to be shorter in the industrial sector than in defense?

Stern: It was hoped to be longer, but it didn't work out that way. In commercial electronics, starting around 1970, there was no product life of any length. The technology moved so fast that you had to get used to recovering your investment over three or four years, or you were out. So in our field, satellite navigation, precision position location, and satellite communications for ships, we just had to get used to the idea that we would have three or four years, and that's it. That had to do with the rapid changes in technology, which then had an impact on the amount of time that companies took to develop. When you knew that three or four years from now entirely new technologies would be available, you could not take three or four years to develop a product. So you had to learn how to develop in a year or a year and a half.

I gave you an example before where instead of two years it took us four years, and that was useless because by the time we came out with this, a new technology had arrived. You had to shorten your development cycle, and you had to develop special techniques to do so. You don't shorten your development by simply saying, "Okay, in the past it took you six months to develop a circuit. Well, now you take only three months to develop a circuit." You had to come up with simultaneous engineering and manufacturing developments that previously were sequential. You had to come up with methods that were unheard of in the fifties or the sixties. You had to pull the whole team together and tell them, "Between such time and such time you're going to do the development work, and your manufacturing and quality people will be involved from the beginning and you will productize as part of the development effort. Manufacturing will be in the picture from day one, and they will provide instant feedback, and then you are going to be able to produce it." You have to have all of them—design engineering, software design, production engineering, manufacturing, and quality assurance—in there together right from the beginning to make sure that when the engineer uses nuts and bolts, that the manufacturing people say, "Yeah, we can use those nuts and bolts, and they are in the right spot."

Aspray: So they don't have to completely reengineer the product.

Stern: That's right. You had to do the electrical and mechanical things in parallel that were historically done in sequence. That really changed our entire development process and our decision process. The decision process was very strongly affected by it because many management decisions, which were usually done a year later, had to be done right away.

Aspray: Did you have to retrain management staff because of that?

Stern: Yes. We had quite a few people who were able to adapt.

Aspray: But, for example, did you decrease the number of layers of management?

Stern: Oh, yes. First of all, we went over to what at that time were somewhat less conventional methods of development. The defense industry had these for a long time because defense systems were so complex that it became clear relatively early that rather than organizing into engineering, manufacturing, and so forth, you had to have product teams to put them together. The concept of project managers was developed in the 1950s, where the basic unit was the project rather than functional organizations: engineering, manufacturing, and so forth. This became mandatory in the industrial area also, to put together a project team that was self-sufficient and competent to do everything that was necessary. They had some consultants from the outside because they couldn't get the best people for every one of the teams. People like Charlie Cahn, who was our chief scientist, worked on many project teams. But fundamentally there was a project manager, and he generally had his project instructions. The progress was measured against those instructions. Of course, the instructions were not immutable, but they were usually followed meticulously. I told you before what happened when I didn't look at something; it was a catastrophe. Incidentally, the distance between success and catastrophe is very small. One or two minor mistakes can cause the difference between success and failure.

In the particular case that I mentioned of our third-generation satellite navigation system that didn't succeed, what really caused the failure was an improper interpretation of a written instruction. The instruction clearly pointed out that the major objective of the new development would be reduced recurring costs. Specifically, reduced manufacturing cost. The second objective was reduced development cost. Now, when you give instructions of this type, they must be interpreted intelligently and loyally. They are clear. That means you may spend a little bit more for development if it saves you on the recurring manufacturing costs. The third objective was the smallest possible size and objective number four was special features compatible with the first three objectives. That priority list was neglected. For the sake of new product features, expensive specialty components were developed. Huge costs. It therefore contradicted priority number one.

Aspray: I see.

Stern: I recall, on the other hand, another case where a similar thing happened, but it was caught. That must have been around 1978, the development of our first high-precision position survey instrument. I caught it in the early design phase. I just said, "You start all over." But it was within a few months from start, and there was time to start it all over. Engineers all have the tendency to go for all kinds of product features and embellishments, so-called bells and whistles. Particularly in the defense industry. I feel it is going to be very difficult to change the defense industry to commercial products, as people have been discussing, because there are major cultural differences. You can't change a culture easily. If the culture of your engineering organization puts features and size onto a higher emotional priority than cost, then they are going to fail in a commercial market. I'm not trying to generalize in some nasty way but engineers will have a tendency to do as much as possible for technical characteristics which are not essential. That is their pursuit of beauty.

Aspray: In this defense environment does that also apply to your manufacturing engineers, who might be more concerned about things such as reliability than about cost in manufacturing?

Stern: In defense everything is very costly. In the commercial business, things are conceptually pure. That doesn't mean that you are not making mistakes. But they are relatively pure. You know that in a certain business, the customer will pay only so much for a feature, and that fundamentally this customer wants to do something specific with this equipment. The thing must be delivered for that specific purpose at the lowest possible price with the highest possible quality. The commercial customer almost never falls in love with fancy technical features. There is no such thing. Again, I'm saying there isn't, but of course there is. In the military business it was a multivariable decision. First of all, the manufacturing engineers generally were not accustomed to the idea of lowest possible cost. But even if they were, the individual customers—and the individual customer frequently was an engineer—decided many things. If the individual customer wanted to have a certain feature, even though it upset the apple cart and cost more, he got it. They knew that it was going to result in substantial cost to the government that was unnecessary. They knew that it would reduce our profit in cases of fixed price contracts. They knew that in some cases it was crazy. But we would do it anyway because that customer controlled much of our income. Being friendly with him was important. Nothing pure. In the commercial business, everything is very pure. Again, there are exceptions.

Aspray: The examples you've given from your industrial products business were ones that were based upon technologies developed for the military. Was that true in every case in your business?

Stern: That was our predominant experience, because that's what I chose at the time I accepted that failing product development in satellite navigation. I felt that taking military technology and applying it with appropriate changes to the commercial market would give us a special position and a special opportunity. So that's what I did.

The interplay between military and commercial products has really been the substance of my career, and it's inevitably somewhat narrower than it could have been. But to go back to earlier years when I was very active in solid state, we applied solid state in military programs at a time when there would have been absolutely no economical way of doing that in commercial products. So I think that there was a great contribution made by the U.S. military to the advancement of technology, both in my organization and other organizations—obviously I was only a small cog in this huge system.

This military-commercial arrangement was both good and bad. It's good in the sense that the U.S. military became the sponsor of technology at the time when others probably could not have sponsored it for a variety of reasons. When the U.S. military sponsored semiconductor technology and integrated circuits technology, other technologically advanced countries were recovering from World War II and they didn't have the means to do any of that. Look at the publications. In the 1950s and early 1960s, we were competing for being first in this and first in that on an individual basis in terms of publications. All of us worked in the U.S. There were no significant publications originating in France or Germany or Japan. There were a few in the U.K. So my only concern at GE in the fifties, when I was in a high publishing mode, was what does that person at Stanford do, and what does the other person at MIT do, and what do the people at BTL or at the RCA Research Laboratory in Princeton do? I wasn't really concerned what the people did in Germany or France or Japan.

We published the first paper on transistor AM broadcast receivers in the January 1954 issue of *Electrical Engineering*. There was a race to get it published because we at GE knew that RCA in Princeton was working on the same thing, and it was a question of where to publish: in the *Proceedings of the IRE* or *Electrical Engineering*, which was published by AIEE. I made the decision to submit it to *Electrical Engineering*, where we would have a better chance of an early publication. It was a good bet. But within weeks RCA had the same thing. Building an AM broadcast receiver, even though transistors were not very good, was not that much of a miracle. The interesting thing was, and at that time we didn't know that this would soon become a common pattern, neither GE nor RCA built the first commercial transistor radios. Instead, it was a Japanese company we had never heard of. Their name was Sony. They are quite well known today.

Aspray: Do you want to make a few additional comments about this great hope today about moving from a defense industry to a commercialized industry? I know you have made the comment about corporate culture before. Do you want to say more about this?

Stern: I think that most of that is politicians' talk. Politicians from the President of the United States down can talk about so-called *conversion* from military to commercial, but to do it is really a different story. If you look at the term "conversion" itself, it sounds and is simplistic. Maybe I'm too literal, largely because I have a fairly strong Latin background. I was asked to go to the Soviet Union a year and a half ago to help them in military to commercial conversion. They wouldn't even be able to convert from inefficient military to efficient military. But to speak of converting from military to commercial operations . . . that's dreaming. It is understandable that large military factories which have a lot of floor space and a lot of employees, for the sake of maintaining reasonable social conditions, should want to use that space and whatever machinery that can be saved in some useful manner. It is the essence of our public responsibility, to make people reasonably happy, and to keep them usefully active. But to say we are going to take this factory and convert it from military to one that produces competitive commercial products just cannot be done.

Aspray: Do you see a solution to this political problem, or the problem that the politicians are facing?

Stern: We are getting into the area of social conditions and political predictions. I believe that we are looking at difficult times to come. I think that the strong reduction of the defense industry in this country is going to have a permanent effect on us, and it's not going to be easy to replace. We have thousands and thousands of highly skilled workers and highly skilled engineers—technical personnel with magnificent technical backgrounds who have become permanent surplus. We have a situation in my state, California, where people are moving out. For the first time in the history of California, people are moving out rather than moving in. Some people will say that's not true, that others are coming in. Most of the people who move in come from Mexico, Guatemala, et cetera. So maybe the total population of California is going up, but the standard of living is decreasing. A lot of valuable people are leaving because there's no work for them.

A similar situation exists in Europe, although the causes are a bit different. Today we are paying for the leadership that we have given to the world in resisting the Soviet Union. We have led the so-called free world in providing a counterweight to the Soviet Union, and we have spent a lot of our budget, a lot of our resources, on defense. Meanwhile the Japanese, for example, have hardly spent any and said, "We will let you defend us. And you are going to pay for it." Today, here we stand. We have a technically brilliant but politically antiquated and largely useless defense industry while they have a magnificent commercial industry. I went into retirement at the right time.

Aspray: I see. New challenges for new people.

Stern: Yes. It's partly challenging, partly tragic. The situation requires a lot of patience and thinking that is really new. Many people in this country—certainly the ruling set of the country—are not prepared for it. It may require conclusions like this: That in the past you saw that everybody who demands a reasonable living must work for it, and that somehow the economy will provide the work. So that when the economy is in high gear, everybody who wants to have a decent living will find decent work. Occasionally there will be low points, and somehow we will get over them. We never really said, "Why should people not be entitled to a reasonable living at those low points." We always got out of the valleys and scaled the next peak. We may have to make up our minds that people are entitled to a decent living, regardless of whether they work or don't work because there never will be enough work for everybody in this society again. But try to explain that to a righteous Republican, or even the many Democrats. I'm convinced that we have reached that point. I think that if we don't do that, we are going to have an increasing number of homeless, an increasing number of beggars, an increasing number of college graduate destitutes, seething inner cities and more and more crime. I don't expect to be around to see what the world is going to do about it. But you will.

Aspray: You said earlier that the IEEE is not as valuable an organization for the young engineer today as it was before. Can you explain that?

Stern: I'm not sure I can explain it.

Aspray: Could you elaborate then?

Stern: I will say this. At the time when I was a young engineer—I came to this country in 1951—one of my first professional actions in this country was to join the IEEE. It was the organization to belong to. It was the organization which had the right publications. For an electrical engineer, it was the organization which from a social point of view was the organization you wanted to belong to and you wanted to be part of. In turn, it was an organization that satisfied most needs of the young engineer at that point. Let's not forget that at that time a very large fraction of young graduates, somewhere between a third and a half, worked in the defense industry. R&D was very important for a large fraction of IEEE (actually its predecessors: IRE and AIEE) and for a large fraction of engineering throughout the country. In the engineering groups to which I belonged in the 1950s, the vast majority were members of IEEE.

Toward the end of my career, due to my personal loyalty to IEEE and because of my experience with IEEE, I tried to push my people mildly into being members. I didn't succeed. By the end of the 1980s, to many of our people IEEE had become irrelevant. Management pressure was of no use. I don't know what the story is numerically, but my conviction is that only a small minority of engineering graduates—electrical engineering, electronics engineering, software engineering, computer engineering graduates—are joining the IEEE.

Aspray: What is it that's lacking in what the IEEE has to offer now?

Stern: I don't think I can answer that in any satisfactory fashion. I did not do a thorough analysis. I would say that, from what I have picked up, IEEE is an elite organization for the older people, for the R&D people, for the people who have international interests. But for the typical engineer, who is doing everyday work in software, engineering design, marketing, or particularly manufacturing, IEEE is irrelevant. Even though there are IEEE societies that are marginally involved in practical areas, it's still considered an R&D type organization. That plus the deteriorating economy has had a very negative impact. Having put so much effort into IEEE as president, officer, board and committee member, from a personal point of view, I deplore the situation greatly. I remember the first engineering group I belonged to in GE. Everybody was a member of IEEE (at that time, IRE). That's why as a young man I became an associate member of IEEE even before knowing why. But I also know that in the 1980s when as company president I put a lot of emphasis on IEEE membership, people talked about it. "Will I be penalized by the old man if I don't join IEEE?" As the Romans said, *Tempora mutantur* . . . times change. And they will continue to change. I hope that on balance the changes will work out for the good of most people. That will take time, a great deal of time.

Chapter 12

Erwin Tomash

About Dataproducts

Dataproducts, producer of high precision printers, was founded in California on March 31, 1962, by a small group of former employees from Telemeter Magnetics and Ampex. The company's founders designed Dataproducts as a high-volume OEM [Original Equipment Manufacturer] supplier with the ability to provide a full range of peripheral products and services to the computer industry. Many of the world's largest computer manufacturers have long been Dataproducts customers.

Dataproducts introduced its first line printer in 1964. Even though the company sold only forty of these printers, early sales figures, for the most part, were encouraging. Two years later Dataproducts established Core Memories, Ltd. in Dublin, Ireland to enter the growing European market.

In the late 1960s Dataproducts introduced the 2310, an eighty-column drum printer that gave the company the momentum it needed to ride the crest of the swelling minicomputer market. The Dataproducts management team decided in the early 1970s to make large investments in engineering, marketing, and manufacturing, thus generating strong OEM growth. By 1973 Dataproducts was second only to IBM as a manufacturer of line printers. The company grew to over three thousand employees and enjoyed unprecedented growth in sales.

Dataproducts enjoyed continued success in the 1980s with its expanded product lines, including line matrix, dot matrix, laser and LED nonimpact designs, and solid ink printers. Distribution was expanded beyond OEMs to include distributors and value-added resellers. The company entered into an alliance in the 1990s with Hitachi Koki to strengthen its financial and technological base. Currently, extensive sales, marketing, and manufacturing networks are in place and Dataproducts printers and printing supplies are in use in over forty countries.

Dataproducts continues to make printers for nearly all of the major computer system providers in the office products field.

Erwin Tomash

Place: Los Angeles, California

Date: June 19, 1993

Courtesy of Charles Babbage Institute, University of Minnesota

Aspray: Perhaps we can begin by having you recount your career, starting with your education and going through your work experience.

Tomash: I was born in St. Paul, Minnesota. I attended the University of Minnesota, which is located next door to my hometown. I graduated high school in 1939 and went right on to the university. I received my degree in electrical engineering in 1943. Because of the war, the courses were accelerated courses, and I graduated in the spring of 1943. In those days, the study options in electrical engineering were limited to power and radio. There were no other choices, and I selected the radio option.

Immediately after graduation I went into the Army Signal Corps. I was commissioned as a second lieutenant and sent to radar school at MIT and Harvard. I spent about six months there. Radar school was an eye-opener. It was my first introduction to more modern electronics. It was also a very good experience. I found I was able to compete quite successfully with the hundreds of young engineers from schools all over the country who were also going through these courses. It gave me confidence in my education and was the first indication of my ability to perform as an engineer in the field. I was in the army until 1946 and spent most of my time in the European Theater of Operations. I did nothing particularly remarkable, though I was awarded a Bronze Star. But I'm not truly a combat war hero.

On my return to the United States after I was demobilized, I returned to the University of Minnesota planning to get a master's degree and also to be an instructor. The GI Bill was then in force, and the university was swamped with returning GIs. They were eager to find graduates to help them. However, I learned very quickly that I didn't care for that and left the university after a couple of quarters. I then took a job with the federal government in Washington, D.C. at the Naval Ordnance Laboratory (NOL). NOL was then in the process of moving to its new home at White Oak, Maryland from its wartime location at the Naval Gun Factory near downtown Washington. I quickly found that I didn't like working for the government. I didn't like all the routine paperwork, and I didn't like the measured pace of the work.

Within a year I left and found a job with a small company called Engineering Research Associates. The company headquarters was in Arlington, Virginia and to my surprise I found that they also had a factory in my hometown of St. Paul, Minnesota. Once I got security clearance, I discovered that the company was building electronic computers. I settled right in. I liked the work from the start and I spent long hours on some of the early vacuum tube electronic computers, which were then being designed to be used for cryptographic purposes. I stayed with ERA in Washington for two years and then transferred to St. Paul. During that time at ERA, I learned a lot about myself, my capabilities, and my inclinations. I found myself drifting towards the generalized rather than the specific. I found I didn't enjoy and didn't excel at circuit design and detailed component evaluation, and other parts of the nuts and bolts of design engineering, perhaps because I was not very good at it. I did find I was quite good at system design, problem analysis, logical design of machines, organization of projects and so on. I soon became a project engineer and then a troubleshooter on a variety of programs and projects.

I also found that, unlike a number of my peers, I was able to express myself pretty clearly and describe what we were doing to outsiders. I seem to have a natural tutorial inclination. I don't enjoy delivering a prepared lecture but I can give a talk or make a presentation from just a few notes. The ERA management called on me often to do just that, which ultimately led me into marketing. In the 1951–52 period I was a project engineer on a major computer development known in-house as Task 29. This later became known as the ERA 1103 or the UNIVAC Scientific computer. I was asked to make a presentation to Remington Rand management on this Task 29. It was actually a *fait accompli*, a hidden and unknown bonus they had acquired when they bought Engineering Research Associates. In the presentation, I suggested that Task 29 be commercialized and turned into a commercial product because we felt it was superior to anything IBM had at the time. Remington Rand accepted this proposal but on condition that I be assigned to help to sell it. I agreed and in 1953 I moved to Los Angeles to open the first West Coast office for Remington Rand Electronic Computers. My choice of location was excellent and the new office was quite successful until it was bought by Sperry in 1955. I left not too long after the purchase by Sperry because for my next career step Sperry Rand wanted me to move to headquarters in New York and become a full-time marketeer. I really didn't want to do that. I didn't want to move my family. We liked California and wanted to stay here.

So I left Sperry Rand in 1956 and joined a small leading-edge company in Los Angeles that was designing add-in core memory systems for computers. Core memories were then very, very new and they had not yet been incorporated into any of the major product lines. The company was called Telemeter Magnetics. I joined it to head up marketing. It was a very small (less than 25-man) company. Within about six months, the president quit and I was asked to replace him. So without any experience in general management I became president of a small, high-technology company.

Telemeter Magnetics (TMI) was owned by Paramount Pictures. How they had gotten into this is a long story and not worth going into here. Suffice it to say that I worked hard to build the company and in a few years TMI was a nicely profitable company doing seven or eight million dollars' worth of business. It had become one of two leading independent suppliers of core memories in the country. In 1959 TMI went public, that is, we sold stock to the public—what is today called an

IPO. A year or eighteen months later—in 1961—Paramount decided to sell its remaining interest, which was still a controlling interest. We ultimately merged with Ampex Corporation. So Telemeter Magnetics became the Ampex Computer Products Division and I became a vice president of Ampex.

Soon afterwards, Ampex itself went through some business difficult times followed by a management shake-up. Within a year, the people with whom we had made our deal were all gone. A new president was brought in to run Ampex and he had completely different ideas. I soon left Ampex. In 1962 I started Dataproducts Corporation. Six or seven key people all from the original Telemeter Magnetics joined me. We decided not to go into the core memories business because Ampex and others were doing a good job in that area. Instead, as Dataproducts Corporation, we planned to supply a variety of peripheral equipment to the then rapidly emerging independent computer industry. We visualized that we would produce a whole set of different product lines. One would be printers. Another would be disk drives. A third would be punch-card equipment. A fourth magnetic tape units, etc. We had a long list of possible products that we thought we could supply to the industry. We did indeed follow that strategy and in the next few years built the company in conformity with that vision.

After about five years we had four or five product lines going and it also became clear that we were making much more rapid progress in printers than in any of the other product areas we had selected. In particular, we were not keeping pace technologically in disk drives, where IBM had been making tremendous investments. Very rapid evolution in disk drives ensued and we were not able to keep pace. By 1970, through phase-outs and sell-offs, Dataproducts Corporation ended up as a specialist printer company. By then we had changed our name to one word and had become the largest independent manufacturer of printers. We sold our products all over the world. At its peak Dataproducts was about a $500 million company and had about five thousand employees. I retired from active management about ten years ago. Just before I did so, I founded the Charles Babbage Institute. That's the story of my career.

Aspray: Maybe we should go back to Telemeter Magnetics. Why don't you tell me about some of the management challenges you experienced? What were the really problematic things for you?

Tomash: When I came there, Telemeter Magnetics consisted of three or four very bright engineers, several technicians, and a small production crew. It had no product line. It had no marketing or finance departments. It was, in effect, a small project team working on two or three government contracts to add core memories to existing vacuum tube computers. As I said, the company was owned by Paramount Pictures. All the contracts had been underbid and were losers. My challenge was to turn the project team into a business. My first steps were to introduce some marketing concepts and to get the rudiments of a financial structure in place. I knew a little about technical selling and a little about marketing. I didn't know anything about finance. I did recognize we needed a good controller. It was clear that we needed to know what projects cost and how we were doing financially. I've never had a course in accounting or finance. However, I feel that anyone who has had a decent technical education, including the usual mathematics courses, can quickly grasp the rudiments of accounting and also how to read and understand a balance sheet.

One experience that I like to mention is the course in electrical measurements that Otto Schmidt taught to engineering students at Minnesota. His second lecture was entitled something like "Beware of the difference of two large numbers." In it he pointed out that small percentage changes in either or both of two large numbers can make a gigantic percentage change in their difference. Of course, the standard accounting profit and loss statement represents the difference between two large numbers. Income is one large number and expense the second large number. The difference is the profit or loss. I soon came to understand that a very, very small change in either income or expense can generate a large swing in the profit—the measure of performance.

My basic challenge at Telemeter Magnetics was to set a marketing strategy and thus determine a course for the business. I soon determined that we needed a product line in addition to the contract work. What TMI had been doing was limited to helping people update their computers. For example, the Rand Corporation had designed the JOHNNIAC memory using Selectron storage tubes from RCA, but then RCA decided not to produce the Selectron in quantity. Instead, TMI built a core memory for JOHNNIAC. We also built core memory for an IAS-type machine [like the computer at the Institute for Advanced

Study in Princeton] at the Aberdeen Proving Ground which was, I believe, called ORDVAC. It was clear to me that add-on memories to update older machines was not a viable business. We did get another contract or two, as there were still six or eight different custom computers in use, but the need for computing power was rapidly being filled by "products" such as IBM 701s or UNIVACs. We clearly were not going to be able to build a business supplying IBM and UNIVAC, who already built their own.

However, there were a number of companies entering the then-emerging computer business, and it did appear as though there was a business opportunity in selling core memories and supplying the cores themselves. Also in providing what we called "stacks," which were the cores assembled in planes called "arrays," and then stacked together to form the main memory component. As a first step, to become known as a leader in the market, I decided on a simple product, a small but complete box of memory that we could sell for ten thousand dollars. Even a small memory was sold for nearer one hundred thousand dollars in those days, and I wanted to impress and indeed define the marketplace by offering a memory in the range of ten thousand dollars. At the time, 1956, transistors were just becoming available as production devices. Transistors were not yet being used en masse in computers, but there were several transistor computers in design. So we decided that our first product should be transistorized as well as low in price.

We settled on a unit with a thousand words of memory. It was contained in a metal box about twelve inches high that could be mounted in a standard, nineteen-inch rack. It included its own power supply and had registers for data input/output and the address. It was as we planned—a black box of memory that sold for under ten thousand dollars. To reduce costs, we did the address selection using a magnetic selection system and ended up with 1092 memory positions rather than the standard binary of 1024. The product was named the 1092BU8. (It had 1092 positions of binary digits each). The BU was meant to indicate that it could be used as a buffer store. That product launched TMI. We sold quantities of tens and hundreds to companies who designed it into their systems. The 1092BU8 established Telemeter Magnetics as a high-quality, low-cost core memory supplier.

Aspray: Who were your typical customers?

Tomash: Collins Radio bought units. They used it as a buffer store. GE was a customer, as were UNIVAC and Burroughs. We also sold units to laboratories. It was used to collect data from punched cards, magnetic tape, and paper-tape readers. It was used to handle the data rate change between slow peripherals and the mainframe. It was seldom used as a central memory. But the product got us enough business to start a production operation. We were encouraged to learn that there was a memory market out there.

Then we got a big contract to do memories for GE for their ERMA system. Those two things, the new product line and the GE contract, gave us financial stability for the next couple of years. By then we had developed a whole range of core memory products, cores, core stacks, and complete memories up to very large ones. At the time Telemeter Magnetics was sold to Ampex in 1960, we were supplying very fast one microsecond access memories to Philco, RCA, and people like that. We sold memories to GE. We sold stacks to Burroughs, NCR, and UNIVAC. We sold to almost every computer manufacturer except IBM. We were really in the business of helping the nascent and budding competitors of IBM. Our challenge was to supply them with products which let them compete with IBM.

Aspray: Did any of these computer system manufacturers other than IBM do their own work in this area? Were you a second source, for example?

Tomash: Yes. We became both a sole source and a second source. Companies just starting into the business, such as Philco or RCA or GE, tended to buy the complete memory answer they needed. Their challenge was to get into the marketplace quickly with a working computer system. If they could buy a reliable peripheral or memory, they did. In the second stage, a few years later, these same companies started to integrate more. That is, they wanted to build more of the system and buy peripherals and components. For TMI this meant selling cores and stacks. For example, Honeywell, which acquired the Raytheon computer activities and therefore had a start into the business, went that route and was a big stack customer of ours, but never bought memories. Very many companies—UNIVAC, Burroughs, and so on—had their own core-making and stack-making operations as well. Yet, they also bought from us. So to them we represented a second source.

Aspray: Did you have any competitors that were independent core-makers or plane-makers?

Tomash: Yes. There were several. One was the company that brought memory cores to the United States. The square hysteresis loop was introduced into this country after World War II by General Ceramics (GC), a company in Trenton, New Jersey. Before and during the war GC was a standard products ceramic-maker which made a broad line of industrial and commercial ceramics. It was owned (at least in part) by German-Jewish investment bankers, the Arnholdt family. They came to America as refugees from Hitler and used General Ceramics as a base to build on here. After the war, a ceramist who had worked with the Armholdts in Germany joined General Ceramics. He had done research in Europe on a new class of ceramics-square-loop materials. Soon General Ceramics had developed and introduced these square-loop materials here. They also obtained strong U.S. patents on the material. They were a major factor in the business and certainly a leading maker of cores.

Core memory, the system using these cores, came from the work of Jay Forrester at MIT. There may be some small dispute over this. There was other early work by Jan Rajchman at RCA and An Wang at Harvard. The engineers at Telemeter Magnetics had worked with Rajchman, and the ceramists, the core-makers, had also come from RCA. RCA, of course, had a long tradition in technical ceramics. We had other competitors. Fabritek, in Minneapolis, got its start as a supplier to UNIVAC and Control Data. Even before the Ampex merger took place, a group of engineers and marketing people left Telemeter Magnetics and started a company called Electronic Memories. They also became a factor in the business. So we had Ampex as a major player as well as Fabritek, Electronics Memories, and General Ceramics. There were also a few smaller ones. Electronic Memories eventually bought General Ceramics. Our breakaway group, a competitor, bought the originator and patent holder.

Aspray: When computer system manufacturers were going out to buy these cores or the core products, what were the most important issues to them? Were they questions of cost, of reliability, of function, of speed of delivery, of reliability of delivery? What sorts of things were of most concern to them?

Tomash: All of these were of concern. But of major concern were performance, uniformity, and reliability. Speed of delivery became important only later. Properly defining performance of the product was critical. Standards weren't yet established and specifications were uncertain. There were complex issues in how to test a core, and what the test indicated. It was unclear how to predict and specify core function within a stack and stack function within a memory. Stack design never did become standardized. So functional specification of core, stack, and memory performance was a critical matter as it related to the desired speed of operation. Reliability was primarily a function of the manufacturing process, but in order to get reliable and uniform performance from the memory system, you needed to start with very uniform cores. The cores had to be alike. Yet these were ceramic products made by an ill-understood process, an "art" that was lost regularly. I say "alike," but [chuckling] of course nothing is truly "alike." Given that every core differed, within what tolerances must we operate? Those were the concerns.

Cost was important. One had to be able to earn a middleman's profit. That is, sell the core, stack, or memory at a price such that the OEM could mark it up to cover his costs and margins and still compete with (be priced beneath) IBM. The cost challenges we faced were in manufacturability of the stack design, and also in the making of the cores, but particularly in the testing. We had to invent and build the test equipment. It was nonexistent.

Aspray: Much is made today of manufacturing techniques and technologies and of quality control issues. Are there things that you can say about these issues?

Tomash: You have to characterize the type of market you're in before you can sensibly discuss these questions. Operating in an emerging market, one that's changing rapidly, is very different from operating in a more mature market. In the former, no standards exist and manufacturing, quality, and process control have different meanings. For example, at Telemeter Magnetics we were shipping products while we were still developing the core manufacturing and testing process. Core technology was not yet diffuse and the processes were very proprietary. In the early fifties, the ceramists, the coremakers, were sort of cooks with tightly held secret recipes and processes. They themselves didn't understand what they were doing, only that this or that worked. Every core plant that I know of "lost the art" from time to time with the result that they couldn't make good cores for weeks or even months. Eventually they would get the art back and make good cores again.

Dataproducts went into the core memory business a few years after we established the company. By then, we were no longer concerned about being sued for using proprietary Ampex information. The reason we entered the core business was that some technical people came to us with a breakthrough idea for making cores more uniformly and at lower cost. We invested in their ideas, which worked out over time.

My point is that at the beginning of a market the issue is how to make the product and make it repetitively. Later on is when process and manufacturing engineering comes in. And even later, when the fundamental technology is well understood, comes sophisticated test equipment, automation, and all the things that are normally done to improve productivity. Most of these steps can only be taken when the product and market have stabilized.

Aspray: What about research and development at Telemeter Magnetics?

Tomash: All of the R&D in all of the companies I've been associated with has been in the engineering area. It has been product development, not research. It has been a constant search for ways to make things, to do something. True research has been limited to the largest companies.

All growth companies do spend a considerable amount of their revenues—on the order of 10 percent—on new product development. This is due to the rapid rate of product change as growth markets develop. Dataproducts always spent about 10 percent of revenue on engineering. We always felt we had to have a product advantage. When I introduced Dataproducts and Telemeter Magnetics to new employees, even in those early days, I said: "We can't get by with 'me too' products. We can't offer products like those of other companies and hope that somehow the customers will buy ours. We have to differentiate ourselves. There has to be a clear, fundamental difference that gives us a cost or performance advantage. We are not going to succeed based on volume. We are not going to succeed because we are nicer to deal with. We have to develop a sound technologically based sustainable competitive advantage."

Hence, for core memories we developed a superior process that was not just different. It yielded more uniform cores. For printers, we developed a print hammer mechanism that greatly simplified mechanical printers and made them more reliable. We sought to apply our efforts to achieving a fundamental lasting advantage. We did not concentrate on "manufacturing engineering" where one takes something someone else has developed and figures out a better way to make that part, but rather tried to innovate orders of magnitude cost or performance improvements.

In the development phases of a market sector real progress is made—real breakthroughs are made—by those who use new approaches to develop componentry and methods to make that componentry, and then design and build equipment based on these innovations. As the market sector matures, volume starts to matter. Dataproducts later on had great advantages due to economies of scale. But we never would have gotten to the point where volume mattered if we hadn't introduced printers that were much more advantageous cost-wise in the first place. Innovation got us into the business. In due course, our continuing advantage and staying power came from the economies of scale, which made it harder for other suppliers to compete.

Aspray: When you were trying to decide at Telemeter Magnetics how much of your revenue to spend, how did you make those kinds of decisions? What were the trade-offs?

Tomash: These decisions are really difficult. Management is under a variety of pulls and pressures. At least in the early days, I was able to arrive at and use a set of rules. We set targets. Our cost of goods was to run 50 percent, engineering would run 10 percent, marketing would run 10 percent, and G&A [general and administrative] expense would run 10 percent. That would leave 20 percent profit before tax to be reinvested in the business. Of course, we never could make those goals. G&A usually ran a little bit more. Engineering often ran more. Also, cost of goods rarely came in as low as 50 percent. As an OEM supplier we really weren't entitled to more than two times markup over cost. We had that lid on our price. One might ask, "Why not just increase the price and have more to spend to do more?" We had to provide a printer to our customer which they could mark up (often double in price) because they were going to have to sell it, integrate it, provide the software, service it, and so on. They needed to at least double our price and still be able to compete with IBM. This meant that our costs had to be no more than one fourth of IBM's price. Roughly IBM's costs. We had to be able to manufacture a wide variety of products at lower volume and still match IBM's costs. That was our challenge.

Aspray: Presumably some of your overhead was less expensive than that of IBM?

Tomash: Oh, yes. Their overheads were higher. I was talking about manufacturing costs though. When IBM used a four times markup, their profits were large—but not four or five times ours. [chuckling]

Aspray: Right.

Tomash: Their profits as a percentage of revenue were only slightly more because they were spending 25 percent for selling expense, 15 percent for administration expense, and so on. Yes, they did have more overhead. As a percentage, bigger companies often have less overhead because of the volume, because of their scale of operations.

Aspray: What was your relationship with your major customers, the DECs and so on? What role did they play in the design of your products, in determining your direction for business? What kind of interaction did you have?

Tomash: In the early years, we had close business relationships, but not in the area of product design. The market was structured by IBM, and our challenge was simply to provide our customers with competitive products of high quality as answers to IBM. Especially in the early years, our product planning was done for us by IBM. A number of our customers had tried to differentiate themselves from IBM but weren't able to succeed. The marketplace kept saying, "IBM is the standard." So the customers wanted products like those of IBM. They wanted simulation. The OEMs turned to suppliers like us particularly for the external peripherals. They limited their concern to the interface. We worked closely with them to make sure that the interface between us was convenient, economic, and nonduplicative.

But by and large they left the insides of our printer to us. Occasionally there would be special needs. NCR, being strong in the banking business, wanted to be able to print OCR characters and the magnetic ink characters—the MCR characters—on checks. That need posed challenges to standard data processing printers, and NCR worked closely with us on that. They concentrated on specifying their need. I recall no instance where our customers' engineers worked with us on the design of the printing machine. They really just wanted a working answer for their system.

Core memory required more technical interchange as it was not a peripheral, but internal. Also there one found different levels of customer sophistication. Several of the customers were capable of designing their own memory systems and simply wanted a reliable second source. Honeywell was a case in point. We did business with them for years, particularly in Europe. We had a large plant in Dublin, and we were their major source for memories in Europe. They just wanted units identical to what they were making in the States. "And don't tell us about improvements." [chuckling] On the other hand, RCA and also most of the smaller companies would come to us for a total answer: "What can you provide? What we need is a faster memory." Or, "We need a cheaper memory." We would then work closely with them using the latest memory techniques. We would design in association with them and try to eliminate duplicated electronics, share test equipment, and so on. Anything to help their speed to market, to make it flow. So, sometimes the relationship was much more involved.

Essentially we remained an OEM supplier. Our marketing was the program needed to maintain liaison with our customers. We did no retail selling at all. Our marketing expense was therefore much smaller than had we been selling directly to end-users. We provided generic literature on our products to the OEM and we often printed it for him. We also offered training classes. We ran a quite large training program for service engineers. They would come to our plants, where we offered classes year round. But we didn't take field responsibility for the maintenance. That was up to the OEM.

Aspray: What kind of technical training did your marketing people have to have?

Tomash: They were almost all engineers. The reason for this was that our true customer was the systems engineer. Basically, we had to sell the OEM's engineering department on the superiority of our product. Their purchasing people would become involved only at the end. In later years, there would be more competition between printer companies. The engineering department would sometimes say: "Both these machines are acceptable and will work." But usually we were able to presell the engineering department on our approach, and we were selected as the supplier. Then the purchasing department would come in, and they would negotiate terms and conditions.

As the company matured, the product line and the market broadened. We started to see printers used by communications and instrument companies. We then started using less technically oriented sales people who were not computer systems engineers by training. But we never used these people to deal with major customers like AT&T or DEC. At these large accounts, one really had to gain the confidence of the engineers. For larger systems deals, it might take two years to do one sell. But then, of course, that order might go on for three to five years.

Aspray: What about the engineers you hired? What kinds of backgrounds did you want them to have?

Tomash: From the technical training point of view, we were quite content with the output of any of the better engineering schools. We did a modest amount of recruiting each year, and we did a modest amount of connection-building with the faculties at UCLA, USC, and Cal State Poly. In general, we had great difficulty adequately staffing our engineering departments. We simply couldn't afford routine, noncreative engineering. We couldn't afford to do what is often done in major product programs, such as big government projects or major company programs—throw manpower at the problem. We expected a lot from our engineers.

I was an engineer by training and had a good feel for both market and product needs. But I did not have a good feel for detailed electromechanical design. Another of our founders, our chief engineer, the person who headed all our development engineering, was an outstanding and inventive designer. Fine engineers are very demanding people and are very seldom good managers. They tend to overwhelm, not develop others. They don't seem to be able to bring them along. It was a constant challenge for us to build quality technical teams that could get the job done. That was our experience in general. In particular, we found it almost an order of magnitude harder to get mechanical and electromechanical design done than to get electronic design done. The difficulty in electronic design lies in development and production of the components, not in the systems design, represented by the plug-in boards.

Aspray: I see.

Tomash: Of course, one can turn the designing of a board into a problem. It can be done poorly. But doing it well is not that difficult. For example, the proper design of the electronic system for a printer, as in the design of a computer, is based upon selection of the right components. The complexity and difficulty is in what Intel does. The next step is easier. It's how you put the components together in a system. Logical design and software design do indeed have creative aspects. But engineering of the hardware of purely electronic systems is not a very difficult thing to do, as you can see from all the clones on the market today.

On the other hand, electromechanical design is extremely sophisticated. For the most part you cannot buy components; you design the components. So in this section you do have all the complexities that are inherent to what Intel does. Designing something that is low cost, that can be made, and is reliable is very challenging. There is an oft-repeated joke about mechanical engineers: A prototype is designed and built and during testing one of the bearings breaks down. The design engineer decides, "We need a heavier duty bearing." During the next test, the heavier duty bearing holds up at which point the shaft breaks down. [chuckling] They replace the shaft, and the next time the bearing goes again. The solution to mechanical problems, things that wear, that encounter friction is often subtle and complex.

Indeed, the reliability and cost breakthrough that Dataproducts made in its printers was due to its hammer mechanisms and the fact that we got rid of clutches and solenoids. A solenoid is actually a quite unreliable gadget. It has that clapper that moves and closes a circuit. It arcs in doing so. Over time the arc burns the contacts and that changes the distance to be traveled. So it doesn't close exactly at the same time or at the same point. A group of solenoids which you require to close at the same instant is a problem that haunted the telephone industry for years. This problem has gone away because of electronic switching. At Dataproducts we tried to get rid of all clutches and all solenoids. We tried to get rid of elements that wear and change with friction.

The best part, the most reliable part of our electromechanical design is the part that isn't there. That's a part that is not going to break and will never cause trouble. So we tried to reduce the number of moving parts to a minimum. This meant trying to understand the functions needed in a fundamental way. Doing this kind of design is very difficult. It's an art and is not taught very well. At least it wasn't taught very well in the years that we were hiring engineers. The path we found was through experience and on long apprenticeship. Engineers in this country aren't prepared to spend much time apprenticing. It is too easy to get managerial positions, to advance by working on large systems projects. We had a lot of difficulty finding electromechanical engineering skills. We did find good engineers from time to time and one at a time. We couldn't hire ten at a time, and when we found them we nurtured them.

Aspray: What about continuing education? Did you have any kinds of programs? Did you believe they were necessary?

Tomash: We certainly encouraged all Dataproducts employees to continue their training and education. We had a companywide program. We would pay the tuition for any employee who wanted to go to school. We encouraged people to get their Masters and MBAs. We encouraged engineers and technicians to take extension classes in engineering. We found in general that in purely technical matters, engineers learned on the job. We paid for any books they bought. We also brought in technical consultants. In one instance we thought optical fibers might have advanced to the point where we could use them to replace some electronics. We didn't know anything about fibers, so we brought in consultants to help us. It was that sort of continuing technical education. We didn't send everyone to school to learn about fiber optics. We remained essentially an electromechanical company. We certainly needed to keep up with the latest componentry, and to that end our people did go to conventions, seminars, and meetings where they heard papers. Also our customers were strong in electronics, and we would learn from what they were using.

Aspray: When you were looking for those electromechanical engineers that were so difficult to find, were you likely to look to people just coming out of college, or did you look to other companies for experienced engineers?

Tomash: Our experience was that we were better off hiring people after they had been out of school for a while. Not because they had printer design experience. There were only a few other printer companies and we respected only IBM. But because we found that there's a difficult transition period for engineers just coming out of school. We weren't prepared internally to nurture them properly, to deal with their uncertainties, and we weren't large enough to place them properly. We couldn't afford to cycle them through different departments.

Occasionally one or another of the professors we knew would tell us about a promising engineering senior, and we would hire him. If he was strongly recommended as a potential star, we would certainly make him an offer. Often the professors made their recommendations if they felt he would do better in a small company than in a big company. Otherwise, we found we did better with people we could recruit from the bigger companies who were at least two or three years out of school. They had settled down somewhat and had our idea of what engineering was all about. On the other hand, we rarely looked for twenty years of experience, although we did bring in a few salespeople from IBM or from GE. They helped us because they brought with them knowledge of engineering discipline and organization.

Tomash: How attractive was Dataproducts to a relatively young engineer? Were you at an advantage or disadvantage or was there no difference between going to a Hughes versus a Dataproducts?

Aspray: I think we were able to compete. Salary and benefits were competitive. We were relatively attractive because we offered the individual engineer more control over his activities. They also were much closer to the decision center. They were aware that the work they were doing was important to the company. There were not several departments between them and the chief engineer, but perhaps only one person between them and the chief engineer. [chuckling] Typically they were assigned a piece of a project and dealt with him directly on it. That was attractive. We were considered a good place to work. We were attractive to people who do not like to work on large-scale weapon systems and aerospace projects. We didn't have big rooms full of engineers' desks remote from the factory. Engineering was much more connected to the life of the company. So we were able to compete for people.

Aspray: You alluded earlier in one of your answers to the fact that engineers often didn't make good managers. How did you solve your middle management/project management problems? How did you staff those positions?

Tomash: Yes, we really did have a lot of difficulty with that. I can't say we ever really solved it. The best I can say is that we settled it from time to time. [chuckling] We kept living with it. By the time the company was over $100 million in revenues, we had a more formal engineering department, a system of project reviews, and so on. By then engineering was certainly more structured and more organized, and we had technical people who viewed themselves as managers, not as engineers. I would say that after about ten years we had a group of very senior engineers, probably numbering half a dozen, who truly were creative engineers. Design engineering is what they wanted to do. We paid them well, and they had implicit authority, but they were not product or program managers. They actually continued to work on technical issues. There was the electronics guru, and the electromechanical guru, the optics guru, the design-for-production guru, and so on. We didn't try to turn them into managers. We had a set of program and product managers who ran projects.

We never really successfully solved the technical management problem of completing an advanced, innovative design and introducing it into volume manufacturing without great expense and delay. We tried many approaches but it remained a very difficult problem for us and we never satisfactorily learned to do it. Introducing products based on new technology into manufacturing is a major revolution more than a transition. A new version of an old product is easy. But a new approach utilizing new technology at the component level—that is a very difficult transition. I don't know of anything to do but work at it. [chuckling]

Aspray: What about in your own role as a top manager? What kind of technical knowledge did you need? How did your background as an engineer affect the decisions you made? Did you feel you had the kind of technical knowledge to make decisions? Or did you have to rely on a certain set of people for that kind of advice?

Tomash: Particularly for companies in newer developing markets, such as biotechnology today or the computer business in the 1950s and 60s, where there are as yet no standards, I think top management must have technical training. You must be able to understand what is going on in your field. You don't have to be a top-notch designer and you don't have to be a technology specialist. If the use of fiber optics is proposed, you don't have to take a course in fiber optics. You do have to be able to understand and judge what people are telling you. You need sufficient technical background to evaluate the business and program trade-offs, the risks. You are dealing with the future where it is easy to generalize and hard to be specific.

For example, there's no question that our increasing population will ultimately create a protein shortage in the world.

Plankton in the sea is a great potential source of protein. Therefore there's no question that plankton is going to be eaten and used as a source of protein by mankind. Now the management question becomes, "Should we open a plankton-canning factory?" To make that decision management must judge and understand all aspects of the situation regarding plankton. These are technical, marketing, financial, regulatory, competitive, et cetera. One cannot decide based only on the generalizations. One doesn't need to be an engineer to understand the generalizations. But one does need technical understanding in order to make judgments about the likelihood of success. In new, immature businesses, there isn't anywhere to turn for that.

My example may be poor in the sense that canning of seafood is not a new, immature industry. Today, the president of a canning company does not need to know a great deal about canning processes and canning machinery. Canning of food has been around a long time. The experience and knowledge base is diffuse and there are lots of sources of information—experts and consultants and so on. But when you're building a business in a new developing market that's where you need sound technical judgment. That is why start-up companies need key executives who are technically trained. Others, such as lawyers or marketeers, can take a company to the next stage or the third stage as the technology and market settle down. At that point the judgments can be based on established norms and general good business practice.

One of the most fundamental questions that torments management of every growth company is how to balance the future with the present? Do you operate to maximize profit in the short term? After all, in earning a profit you are creating a healthy, strong, more secure, financial base. How much of the profit should one spend on engineering of new products and thus create greater future strength? I tried to answer such questions.

The best answer I ever came up with on this was that we had a number of different constituencies to satisfy and that the answer was somewhat different for each of them. I started with the questions: What are we trying to do? That is, who do we work for here? I came up with five constituencies that I felt we were to serve. Their needs differ, their desires differ, and their priorities differ, but still we had to serve each of them every day. Customers were our major constituency. Next was our shareholders. Third was our employees. Then, we also needed to support our vendors and serve the communities that we were in. Each of these viewed the company, Dataproducts, through a different prism. Your customers are not concerned directly about your shareholders. They want you to stay in business. Your shareholders should be concerned about the customers, of course. I could continue down the list. So what is management to do? Do you make as much money as you can now and return large dividends to the shareholders? Do you invest all profits into R&D and thus return little now but much more to the shareholders in the long run? That is, if you last. So management must try to balance the short- and long-term desires of the people they serve. Most of the constituencies would like you to do it all. They would like great short-term results with great long-term benefits. [chuckling] Why not? And great companies sometimes provide these.

For instance, stock analysts are notorious for this. They come to the company to visit, and if you're quite profitable, they quickly ask, "Well, that's nice. But what are you doing about future investment?" But if you're investing heavily they say, "Well, what about your current earnings?" [laughter]

The incentive and reward system that we have in this country is oriented to the short term. This influences management greatly. I feel this is a great weakness *vis-à-vis* the rest of the world. Even so, I've met many people who started companies and very few that did so just to make a lot of money. Everyone likes to make money, but stronger is the desire of the engineers and marketeers to build something. They want to build a company. That's often a stronger drive than making the money itself. Those dollar rewards may come with success but they're not the driving force they are often touted to be.

Aspray: You mentioned earlier on that there wasn't the right kind of test equipment when you first were involved in Telemeter Magnetics. What were some of the other issues about tools for engineering and production that came across your way?

Tomash: Components are probably the best illustration of what I had in mind. If you look behind a component manufacturing business, you'll find that the continuing challenge is in production testing. In order to test something adequately, the tester has to be an order of magnitude better than the testee. So if you are building a high-precision, high-performance whatever, the test equipment has to be high precision and high performance squared. It's a continuing problem as the product line advances.

The ferrite core is a good example. In the early days a core was made with a tableting machine, a machine borrowed from the pharmaceutical industry. The magnetic materials were ground up and mixed in the form of a powder. Then this powder was handled just the way they handle aspirin powder. The tableting machine had a small hopper in which the powder was placed. The hopper would then swing over a die and deposit an amount of powder equal to one tablet's worth of ferrite core. The top of the die would be brought down to form the toroid by pressure. The machine would then eject this formed core. These soft cores would then have to be taken to a kiln to be fired and turned into a ceramic.

There were several processing problems that arose. Core memories require cores of much greater uniformity than this process easily yielded, and stringent production discipline was required. Not to mention the firing process in the kilns, which also added nonuniformity. So when the process is complete and we measure the magnetic properties of ten thousand of these cores that we've just stamped out and baked, we find a wide range of values and characteristics. A number of different things were happening. First of all there's a little bit more powder in one than another. You can lose a grain or two of powder because the machine was just dumping a charge in and around the die. Secondly, ferrite powder is abrasive. It's iron oxide. So the die is getting worn just a bit each time you use it. If you had made ten thousand cores with the same die, you can be sure that the dimensions of the toroid are different in toroid number 10,000 and toroid number 1.

The other important dimension is the thickness of the toroid, and that depends on the pressure used in the die. At first we used a purely mechanical encapsulating machine taken over from the pharmaceutical industry. Later we used a pneumatic system, and we still found that core density could vary over a wide range. That is, some of the cores were packed more tightly than others. Then as they went through the furnace, did all the cores experience the same temperature? This depends on the cross section of the kiln and whether the cores in the middle of the kiln got more heat than the ones on the outer edge or on the trailing edge. Next is the question of cooling or quenching. Did some cores start to cool faster than others? You can see that in any complex manufacturing process like this, there are an endless number of places where the equipment and the process used influence the end result.

Returning to my ten thousand cores, before using them I must do ten thousand measurements. We had to develop mechanical equipment that automatically handled cores: that would select a core and probe it without chipping. The electronic equipment would run some kind of a test on it while it was held and then based on the test fed into a yes or no vial. In the early days, we found that the variations were so great that we had ten bottles to feed into. That is, we tried to get groups of cores nearly alike out of a wide mix.

There are two approaches to this type of problem: one I call the manufacturing engineering approach, and the other the development engineering approach. In the manufacturing approach, the engineer considers the defects of each step in the process and tries to cure them. For example, he puts an improved valve in the pneumatic system so it accurately supplies the same amount of pressure every stroke. Indeed, he senses the pressure and uses a feedback loop to control the valve so that the pressure is constant. He automatically weighs each core after each time he handles it, and so on. He accepts that the core die is going to wear, so he counts usage and replaces dies on a regular schedule. In a manufacturing engineering approach you improve the process step by step through tighter controls and by engineering little artifacts and aids that help. If there is a problem with the heat profile of the kiln and its cross section, perhaps he'll only use the center, or he'll redesign the carrier. If necessary, you redesign a new kiln, one that will redistribute the heat so that it yields a more uniform result. That's a manufacturing engineering approach.

The development engineering approach is to try to develop a new process. It says: I'm not going to use powder at all. I'm not going to use dies that wear. What I'm going to do is make a slurry of all this powder, put in a binder, and roll it flat into a very thin sheet. I'll use two adjustable rollers. I can adjust those very accurately and know that they are well within the tolerances I need. I will have the same thickness and density all the time. I'll also put a lubricant into the slurry, one that will survive the heat of the kilns, and which will oil the dies when they cut. That way the dies have long life because I'm lubricating them. And I've got a kind of a cookie cutter process.

This latter process is indeed what we did. With it, Dataproducts went back in the core business. We set our competitors on their ear because our costs for a good core were one-tenth of their costs. We didn't have to shake any powder. We didn't have to change dies. We didn't use a tableting machine. We cut thousands of cores at once. We didn't have one die; we had whole rows of multiple dies. By the way, we introduced a whole new set of problems that the manufacturing engineers had to come and help us with. But we did solve the basic problems in the powder manufacturing process by developing a way to get around them rather than fixing them. It's a challenge to balance these two ways of doing things.

Similarly, in the area of test equipment we went to two-stage testing. We could quickly measure with multiple-headed testers the rough magnetic characteristics. We could quickly and cheaply get rid of truly bad cores. We didn't know if the rest were good yet, but we knew at least they weren't bad. We got rid of bad cores on which further effort was a waste of time and money—perhaps 5 percent of them. The next step involved a whole new set of automatic core testing equipment. Here the optimum solution was to go the other way, to go to small inexpensive handlers and testers that we could adjust individually.

The correct answer to these issues is, of course, to apply the latest technology in both the test equipment and the process. This kind of thinking was quite traditional in the chemical industry. When we were doing this in the late 1950s and early 1960s it was a new set of issues for the electromechanical and electronics industry. So the net result was that we were able to create better, lower-cost components.

Aspray: You mentioned a group of people leaving Telemeter Magnetics to start their own operation. I suppose one always has to count on a certain number of people leaving, but not necessarily leaving to form their own competitive company. Can you talk about the kinds of issues that you face as a manager when that kind of situation occurs?

Tomash: It was particularly difficult for me because I really have a strong, deep entrepreneurial streak. On the one hand, I felt that we had built loyalty to the group and to the company, and that we treated people well and built an environment for teamwork and strong association. So it really hurt when it happened. I couldn't criticize, much less dislike, others for trying to satisfy their own entrepreneurial impulses. In the 1960s, start-ups seemed to happen mostly when people got uncomfortable with their situation. It happened less because they wanted to start their own company than because they were unhappy doing what they were doing. At the time, there was not too much venture capital around. This may not be a valid observation today. But I left Ampex unhappy and started Dataproducts because Ampex wasn't going in the direction I wanted. The fellows that left Telemeter Magnetics to form Electronic Memories left because Ampex was coming in, and they didn't like what was happening with Ampex, didn't like Paramount selling us out, and so on. In addition, I may have triggered one start-up by a reorganization in which I reduced the responsibilities of one individual and promoted two others. Whereas I had had one number two person, I now had a trio. The one who was previously the number two person left to form another company.

Dataproducts spawned several companies, some of them quite successful, some that failed. You asked me how I handled it. Intellectually, I felt as long as they played fair—and fair, to me, meant that while they certainly took with them the content of their heads [chuckling] (you can't take that out), they didn't do anything unethical; they didn't take their project, their designs, customer lists, et cetera. They didn't bad-mouth us, and if they competed, it was indirectly. Under those conditions, I had a lot of trouble getting up a big head of steam over their leaving. Emotionally, as I said, I was always hurt and disappointed.

In the case of Pertec, which I think was one of our first Dataproducts breakaways, they went into the magnetic tape drive business. We didn't have products in that area. Later Pertec went into the disk business, but we had already gone out of the disk business. They didn't go into the printer business. In another case, the group went into the printer business and eventually failed. They took one of the design approvals that we had set aside as being not too successful, and they pursued it. They also took a few of our people.

In this case, we competed hard to protect our customers from their product. Really, I do think start-ups are a kind of tax that a successful company has to pay, [chuckling] and it puts something back in the well. It's not too bad. It's a process similar to your children moving away and setting up their own place. As long as it isn't stealing, which I define as taking designs, taking customers, and so on, it's a beneficial process. I was upset for a while with the Electronic Memories founders because they did go into direct competition with us. We were in the core memory business. They did try to bad-mouth us. All of this was unnecessary, I thought. As I look back on it, maybe it was necessary for them to bad-mouth us in order to justify their own actions and to make themselves feel a little bit better about it. But all my annoyance faded away after a year or two.

Aspray: Can you tell me about the start-up of Dataproducts? How did you make the decision to go into the business? What kinds of people did you look for to pull it all together?

Tomash: First I had decided to leave Ampex. When I did so, I really didn't have starting a company in mind. I didn't know what I was going to do. But I was unhappy with the new Ampex management, and I had plenty of confidence in myself. It was in 1962, and I was forty-one years old. Immediately after I resigned three of the key people in the company, the chief engineer, the head of marketing, and the chief financial officer, came and said: "Hey, let's start a company." [chuckling] My first reaction was, what are we going to do for money? Most of us had held options in Telemeter Magnetics, and when it was sold to Ampex, all options were cashed in. So among us we had a little money, but we certainly didn't have a lot of money. I started to inquire around. I went to see people. I went to see an investment banker that I knew. I went to see the lawyer for Telemeter Magnetics, who incidentally has remained a strong personal friend all my life. I also contacted my old friends in Minolta. Almost everyone encouraged me. They felt we could raise money if we put together a business plan.

The next step was to decide what we were going to do. We decided we'd go in the computer peripheral business and that we'd serve the systems industry rather than compete with it. We then developed a little theoretical model. We decided to see if we could get a head start by buying a company in existence and so gain a little momentum. I started to look at a few possibilities. I did this myself. I was the only person unemployed. Everyone else was still working hard at Telemeter Magnetics, which by then was a part of Ampex. It wasn't at all clear that we were going to be able to start a company.

Then Arnold Ryden from Minneapolis phoned me. He was a business associate of my old friend from ERA, Bill Drake. Ryden was chairman of Telex and was trying to expand it. Telex was originally a hearing aid company. Ryden had started a division to develop disk drives in St. Paul with some people from UNIVAC. He asked me if I would come up and run Telex. They were looking for a president. I said, no, I wasn't interested in most of what Telex did and the idea of running a miniconglomerate had no appeal. But in the conversation I said, "Why don't you sell the disk drive business to us?" They said, "We can't sell it. We've put a lot of money in it. It's important to our shareholders. We're a public company. It doesn't have much net worth. It will be worth much more later. Besides, you don't have much money anyway." It was true that we didn't have any money at the time.

A few days later. Ryden phoned again with a new idea; he was quite a brilliant financial man. "We'll spin the disk drive division off to our shareholders. We'll create a new company so that there'll be Telex the hearing aid company—the old Telex—and this new company which is the disk drive business. We'll sell you stock in the new company, which will get new money into it. Your group will have major ownership in a new public company." So that's what we did, in effect. We formed Dataproducts, put our money in it, merged with the disk drive part of Telex and then spun it off to the Telex shareholders. That's how we got Dataproducts started. We then raised about a million dollars from three SBICs to get the working capital for it.

Telex had a contract with GE to deliver disk drives, but it couldn't build them. That was our first challenge. Looking around Telex, we also found a printer project. This was located in Detroit, an operation with four or five people. They had offered a printer for five thousand dollars. Printers were then selling for thirty thousand dollars. We hadn't particularly thought of going into the printer businesses up to that time.

I went to St. Paul the day we made the deal and signed all the papers. Things were happening everywhere. We had inherited a union in St. Paul. They went on strike the day after we made our deal. April 1 we made the deal; April 2 they went on strike. I was there to see what we really had bought into. GE was trying to cancel their contract because Telex hadn't delivered. I asked Graham Tyson, one of the key people from Ampex/TMI who joined us, to go to Detroit to see what was there. Two days later he reported. "There's nothing here. There are five or six people in total. They've got one printer they're getting ready to ship to DEC. It has never run. They haven't tested it yet.

They haven't completely put it together yet." Tyson said, "I'm going to shut it down." So I said, "Before you do, are there any good engineering people?" He replied, "I think there are a couple of them that aren't too bad." So I asked Cliff Helms, who was our founding chief engineer, to go to Detroit and interview them before we shut it down. The next day he reported, "Graham's absolutely right. There's nothing here. If this thing ever does go together, which I doubt, it won't last fifteen minutes." But he said, "There's a germ of an excellent idea here for a print hammer mechanism. They're using a coil in a magnetic field, like a loudspeaker instead of solenoids. The way they're doing it, I don't believe will ever work. But I like the idea of getting rid of the solenoids."

The upshot of all this is that we did shut it down; nobody would move. It turned out there was only one engineer we made an offer to, and he didn't come. After we shut it, we started a printer program in Los Angeles under Helms' direction. The basic idea was that instead of a solenoid, which all printers then used, we would use a free-floating coil in a magnetic field. A pair of magnets on either side of a moving coil. It was similar to a loudspeaker, but instead of pushing a diaphragm to make sound, we would push a hammer to strike the paper. This idea was implemented very, very poorly in Detroit, but the basic notion was a good one. On it we built the whole printer business, eventually.

When I started Dataproducts, we were flooded by people from Ampex who wanted to join. But we chose carefully. We picked a key financial man, who really was surplus because Ampex had its own chief financial officer. So we had Bill Mozena as our financial man. We had Cliff Helms as our key technical person. Graham Tyson was the head of Operations. He was also a good engineer. Russ Dubois was the head of Marketing. He had done the marketing at Ampex. That was the team that signed on April 1.

We had to quickly assess the problem with the disk drives in St. Paul. It became evident that they'd made a mess of the electronics and that the logical system was just a jumble. They hadn't really had a decent logic designer at all. So Irv Weiselman, a proven logic designer from Ampex, was hired because we needed someone to sort that out. We also felt the need to get somebody to handle GE in Phoenix. We needed to hold their hand, and keep them in the family until we could get the product fixed. What was doubly worrying was that they could cancel the contract and retain full manufacturing rights. They didn't really want to cancel. They didn't really want to manufacture. But these disk units were essential to their program. If they lost faith in us and felt we couldn't fill their needs, we would drive them into the disk business, disk drive them into the business. What we needed to do was establish and maintain their confidence.

We picked another Ampex/TMI person by the name of Jack Ogg and asked him to join us. I think he was our seventh or eighth employee. He went to Phoenix to keep things calm. At the time, we didn't know if we could fix the problems because we didn't know what was wrong yet.

That was the start of Dataproducts. We ultimately succeeded in keeping GE-Phoenix as a customer. We built a few hundred big disk files for them. They used big, huge platters for disks. They were not as large as this four-foot table, but I think they were thirty-two inches across. It was like the early IBM RAMAC. We ultimately built some even larger units. I saw one of these a couple of months ago in the Science Museum in London. We shipped a huge unit to Cambridge for their Atlas machine, and it is in working order at the Science Museum today. It's on display there. Anyway, the GE contract gave us a start. In the meantime, Cliff Helms came back here to California and started the printer product line using the venture capital we had raised. It took us about three months. Once we got things fixed we had the GE contract for cash. Just as we were running out of money, we started shipping to them. In the meantime Cliff and Tyson started to build a little operations nucleus to work on the printer. It was a very small effort at first because Cliff was just figuring out what to do. He was literally the project engineer himself. He was ideally suited for that challenge.

Aspray: In what way?

Tomash: First of all, he's a very, very good engineer. I didn't say that lightly and I don't say it about many people. He's truly an engineer. By my definition, a person who's really an engineer can deal with almost any technologies, can teach himself what he needs to know, and has a fundamental understanding. Cliff starts with $F = ma$ when he has to and works it out from there. He states the problem, analyzes it, and solves it. He understands technology in the fundamental sense. Cliff also has insight into mechanisms and is very innovative in development.

He came up with the idea that we would build a printer with no solenoids, no clutches, no brakes, and feedback control in the paper drive. As I said, he used the moving coil principle, but he mounted the coil on flex pivots, that is, spring steel legs. The current was conveyed in the coil right through the legs so that there were no coil contacts to make and break. There was no friction wear; there were just pivots and spring legs. So the hammer has virtually endless life. There are Dataproducts printers working in the field that are thirty years old. The hammer mechanisms are still going. Spring steel that's flexed within its tolerances just keeps flexing. He is that kind of an engineer. He tends to solve things in a fundamental way rather than tinkering with them. He sat back, thought deeply, and came up with his solution. Of course the printers five years later had continuously evolved—smaller, faster, simpler, and all that. But the first idea was from his notebook of the first six months. He got rid of the clutches and brakes used in the paper feed by using printed circuit motors, which were just coming on the market. You could use electronic feedback and an optical disk to lock the paper, so you didn't have a clutch and its associated wear.

Aspray: As the printer industry matured, and as Dataproducts got larger, what were the kinds of new challenges the company faced?

Tomash: At the beginning, disk drive shipments carried the company as a business. Our printer product challenge was to get the design done, get into production, and to gain credibility with customers by getting them to test them. There was an established supplier of printers, Analex, in Boston. Their printers were priced at about thirty thousand dollars. They used solenoids and their printers were notorious for needing maintenance and realignment. People regarded Analex printers as successful if they printed pretty well for six or eight hours. After that you had to spend an hour or two to realign and maintain them. But they were the only game in town if you weren't IBM.

We had the challenge of getting into production and getting first customers for an unproven product. We readily had great difficulty to take this new design into production. It meant learning a whole new set of technologies and for the first couple of years we were struggling to get those right. Of course there was the challenge of product costs. You can't control costs when you're first learning. When you don't know what you're doing or how to do it, you can't cut costs. You have to have stability in what you're doing and then you say, "Now I'm going to do it a better way."

Our printer depended on printed circuit motors and depended on our new hammer banks. The hammer banks were by far the more difficult problem. We had three new challenges. New very strong magnets; they were a new type of Alnico which was still not uniform, so we had to take into account individual variations in the magnets. A typical engineer might simply write a tighter spec—in effect, transfer the problem to the vendor. That's all very well, but usually the costs go way up because the vendor ends up selecting magnets. The proper approach is to work within the characteristics of the production material.

The second challenge was inherent in the hammer design. When any print hammers strike the paper, there's a force of nearly 1000 G experienced. In our design this impinges directly on the coil. How do you insulate the coil and keep it from shorting internally. The solution was to use flat wire. We made a sturdy package of flat wire so that the turns lay one on the other instead of crossing and jumbling. The shock, when it hits, tends to travel through the coil. But that meant using a flat wire. We found that we could buy flattened wire. It's made in the instrument industry for ammeters and things like that. It's about a dollar a foot. We were talking about literally hundreds of miles of wire, and we'd like to pay less than a penny. So we developed wire-rolling mills to flatten the wire and managed to do it without changing the tensile strength. Indeed, once we got going and started buying aluminum in quantity, people from ALCOA showed up to see what we were doing. They'd never seen this before.

Third was the problem of coil and contact insulation. For this solution we borrowed epoxy then starting to be used in the aerospace industry. Epoxies were then brand new. We used epoxy to coat and thus insulate the flat wire. We wound the coil and then baked it. It became a solid mass that wouldn't move internally regardless of how many Gs were applied. We used a different epoxy to insulate the connections. I mentioned that the coil carries the hammer and is attached to flex pivots. The current is sent through the pivots. So, somehow, you have to insulate the coil from the rest of the structure.

The reason I describe this in detail is because it only took us another year and a half to develop these manufacturing techniques and machinery to make the printer hammers, to develop the disciplines, and so on. That was a serious difficult challenge.

Our next challenge was getting the product introduced and accepted. We used price. Our price was about fourteen thousand dollars versus thirty thousand dollars for ANALEX. We had to get the attention of the market. Engineers quickly saw the merit of our design. But decisions to commit are based on more than just the design. If you're already using ANALEX, you have trained service men, you have a large inventory of spare parts, you have machines you haven't yet written off, which you're renting out. There are many factors to be considered.

After about a year and a half, we were out of money. That was another big challenge. We had not yet delivered printers in any quantity. We were shipping disk drives to GE, but that was not enough to keep us afloat. We had used all the initial money to fix the disk drive product and to finance the printer development. As is often the case, we stood on the brink of success— out of money. We decided to make a rights offering to our shareholders, as is more commonly done in Europe. We offered stock at a low price and we got our investors to exercise their right to purchase the shares. That got us enough money and stabilized the company. Then our challenge became developing a flow of competitive products.

A few years later we ran into other serious problems as I said in my remarks about the disk drive business. We finally found we couldn't afford the engineering it took to keep abreast in that business. We simply didn't have the financial resources.

Aspray: Do you want to make any more general comments about capital, its availability, and the way it affects the running of a technological business?

Tomash: Yes. I have some observations and experience that I should talk about it. I think the concept of venture capital funds is a very good idea and is very important for start-ups. However, I think that it has matured into something that really doesn't fulfill this function any longer. Venture capital today has become a big business of large funds. It's very difficult for these funds to manage a large number of very small investments. Therefore, they tend for this and other reasons to select only for large opportunities. They want to invest in companies that will be able to go public at the end of five years. In today's markets that means a $50 million revenue company with a chance to be a $500 million company. Therefore, they tend to invest in teams of people, in clearly identified products, and in markets that are already established, not new. Fifty million and 500 million are not start-up markets. The result is overfunding of start-ups in large established market sectors. These don't do the economy or the funds any good.

For example, we managed to start so many disk drive companies that no one has made any money. Seagate did develop, and there are a half a dozen others that may survive, but the venture capital firms had twenty or thirty of them going. There is indeed a big market for disk drives. But by overfunding they made certain that no one would be a big success. Venture funds nowadays put large pools of capital together. They invest a few million each and the deal ends up at 25 million, so it must be a big deal. We need more smaller venture capital firms that have more entrepreneurial direction and undertake higher risk deals such as true start-ups.

It's very hard for an individual with only one or two other people involved to get funding today from venture capital. The venture capital specialists demand that you have a complete management team and a clear product idea. They demand too much sophistication at the outset and demand too much from the lone entrepreneur. Yes, he needs to have a business plan, but it ought to be a modest, not very sophisticated, plan—one that can really be put together by asking some simple questions. He needs half a million dollars, not $5 million. But then he also needs support and faith. In almost every business you go through at least one valley in the early days where it's questionable whether you're going to make it or not. Maybe two. I know that was true of Federal Express, a venture deal that almost shut down. That's when you need the courageous venture capitalist to give it the extra money to push it over the top.

Other countries use government-supported bank capital to finance new ventures. Our banks are just not in this start-up, small company thing at all. I think that's a weakness. We rely on the stock market both for money to grow companies and for the payoff. The public offering is the window where the venture capitalist presents his winning tickets. In Germany, Nixdorf, who was a customer of ours for years, was financed by a regional or state bank who wanted job creation and therefore wanted Nixdorf in their area. They backed him until he could go public when it was a substantial company. That's an example of patient money. The Japanese use their banks for equity money and have more patient money. The difficulty with our method of using the stock market and venture capital oriented to the stock market is that it really does focus management on short-term results. All the rewards and incentives are short term. Some people use the phrase "investing," while other people use the phrase "speculating" to describe the same actions. In any case, you want the shares to go up. It puts an undue pressure on the management of young companies to perform quickly and to strive for quarter to quarter improvement. The pressure is to avoid the best kind of investment, which is the investment in people and know-how and infrastructure.

For example, at Dataproducts, when we were smaller, I could buy a million-dollar machine (if we had a million dollars) and its cost did not greatly impact profits. If it does not become obsolete, you are able to write it off over eight or ten years. The impact on earnings is $100,00 or $125,000. If I decide to spend a million dollars to develop a new product line, engineering and development salaries amount to a million dollars spent and are "lost" immediately. But which would you rather have? A piece of machinery or a new product?

If, for instance, you decide that the thing to do at Dataproducts is to double our marketing capability and open new offices because that's going to give us the channels and the penetration we want and that costs a million dollars, that's a loss. If I bought a piece of machinery, it's an asset. Which would you rather have? A doubled marketing capability or a piece of machinery? So the public market measurement scheme—there's nothing wrong with the accounting *per se*; you have spent the million dollars, you don't have it, it's gone, you shouldn't put it on your books—but according to the measurement scheme that comes with the public market, earnings just went down by a million dollars. Therefore, instead of adding $10 million to $20 million to the shareholder wealth, we have hurt valuation of the company. We've hurt all the shareholders by doing what's right for them. That gets me back where I started—with the long-term/short-term. So what all the shareholders would like is increasing profits *and* investment for growth. That's a very, very difficult thing to do. This tells me we have more companies going public than are able to meet these expectations. There are fewer companies being financed than deserve it. Many companies don't need to be public, and they aren't in Japan or Germany. But they are worthy job-creators, worthy of fulfilling the entrepreneurial urge of the founders and good for society in general. They pay taxes. They do a lot of things we as a society want them to do. They become suppliers to big companies. They make money for the owners. They fulfill an important role. Some of them will merge and become public later. All in all, I think the short-term syndrome is too tough on a small new enterprise.

Aspray: I want to ask a similar, general kind of question about government and its role in these kinds of businesses.

Tomash: I am of a mixed mind there. My experience with government in one aspect is very positive. In my judgment, the seeds for the computer industry were planted by the government's risk taking. The government bought computers when no one else did. Even though some of the early customers were aerospace companies, they were all on government contracts. The government bought all eighteen of the IBM 701s, most of the 1103s, and all but one of the early UNIVACs. Because of their needs and demands they also encouraged a lot of development work which later showed up in commercial products. The commercial pull came later. So the role of the government as a major risk-taking customer, one who has large-scale problems and can afford expensive new technology, is a very valuable function.

In general, I'd rather see the government be a leading-edge customer than see it try to set the direction for industrial or technological development. I think the economy gets helped more fundamentally when the government asks for a network to use and they're willing to help finance it, than when they sponsor network software development and logic. It's the solving of real live problems and creating demand that has pushed our industry along. There I think government has a positive role, and I wish that somehow that role could become more socially acceptable and palatable.

The government in its regulatory role has a very, very heavy hand. This is particularly true for small new enterprises, and it's on all levels of government. You can't build a building in the county of Los Angeles in less than two to three years. California has a workman's compensation system and state laws regarding employment that are outrageous. Federal OHSA, EPA regulations are no better.

We pride ourselves on all the jobs we've created in the services sector. In my opinion, a significant fraction of these, those that are not minimum wage, are nonproductive and have been built through government regulation. The tax law is an accountants' and attorneys' retirement act. The public pays for people who are going to protect them and help them, and facilitate for them against another set of employees called civil servants that they are also paying for. With racial, sexual, age, and gender discrimination laws, it is a wonder small industry is willing to risk hiring anyone. No wonder the "temporary" employee business is booming. The challenge is to maintain a balance. We want a fair, equitable society, but no one feels we will get there by government regulation.

I don't know if I would even consider starting a company today. I was an acting CEO for a little while recently. It came to me as a shock to learn that as a practical matter you can't just fire somebody for incompetence. You must first build a case, and that takes at least six months to get the file right. What does this do? It prevents you economically from replacing that person for at least six months to a year. You continue to suffer the costs of that incompetence. You lose the time. Then finally you do replace him. I suppose we have prevented abuse, but at what cost? For whatever percentage of abuse we had, is it as large as 3 percent or 5 percent? We've spread this cost on everybody to prevent this abuse. The government's hand, that indirect hand, in regulation and in an overlegislated society is very costly. It introduces distortions and aberrations, and makes it very, very difficult for anybody to start a company or to carry on a business.

Aspray: What about an international trade role of the government?

Tomash: I think the government is right to try for a level playing field. However, the mechanisms that are available for protection are a joke for small business because they are so expensive in both time and money. If you really feel that you're being discriminated against and you'd like to bring a modest, straightforward case to the Federal Trade Commission, it can cost a hundred thousand dollars a month in attorneys' fees. The process we have requires you to prove damages. But to prove damage takes a few years. By then you're out of business.

If I may generalize for a moment, in European countries where civil servants are not looked upon as potential enemies, they in turn don't have an adversarial attitude towards industry. In those countries, civil servants are expected and encouraged to exercise judgment. By contrast, our system does not reward risk taking and encourages the avoidance or use of judgment. It tries to substitute rules for judgment. If I could show that this glass here is being sold in Japan for fifty dollars and is being sold here for five cents, and I could bring you the sales slips to show you that and a picture of both items, it could still take years and millions of dollars to get a court order dealing with that. No one in government will make a judgment until they get a court ruling. No one will decide anything major short of a court decision. So you see by emphasis on process and regulations we simply handicap ourselves and slow our whole response mechanism.

I think that American companies can compete well abroad on a level playing field. Even without a level playing field they do and should be encouraged to do it. Not that they need much encouragement. The products that sell are the items that are attractive because they represent leadership. We sell jet airplanes because other countries don't make them as well. We sell ultra large, ultra fast computers, and so on.

Aspray: In the printer business Dataproducts no doubt had Japanese competition. As I understand it, Dataproducts now has a Japanese parent company. Do you want to talk about doing business in competition with the Japanese?

Tomash: Yes. Dataproducts was able to compete in the OEM market very successfully against the Japanese. They tried to compete and later told us they couldn't make any money selling printers against us. We sold worldwide on our own. We offered quality products at good value. The market was relatively small, and it was only a niche market. Dataproducts had perhaps 40 percent of the business, and it was a $500 million company. Relatively speaking, it was a small business in terms of worldwide markets. Where the Japanese excel is in mass production of standard products, and there we couldn't do as well. We didn't think mass distribution or mass production. So Dataproducts missed the personal computer wave, and Hewlett Packard caught that wave. Hewlett Packard showed how to join with and compete with the Japanese. Their two best selling products are based on technology. The case is made in America, and the electronic board is made somewhere off-shore—not Japan. But the Canon engine inside is what makes that printer, and that comes from Japan.

The Japanese have excelled at mass production, in volume, particularly on smaller units sold to the end user. We didn't do well there. Whether we could have done well, I don't know. I do think Dataproducts as a company could have done better in this last decade than it did, and I fault the management. Dataproducts didn't try to lead in nonimpact printing. It stayed too long with the hammers and its established technique. It let the next wave catch them. Today, we may think that laser printers are the cat's meow, but I assure you—although I don't know what it is—something will come along to obsolete lasers. The new technology will be cheaper, better, perhaps provide color, whatever, but inevitably something else will replace lasers.

There are several disadvantages intellectually, from a technical point of view, to the laser printer. It's not a direct printing approach; it's indirect. Then there is the need for replacement of the cartridge. The process has a great deal of friction inside the cartridge. One reason they went to a replaceable cartridge is that toner is abrasive. You need new parts to account for the wear. Intellectually, there's room for someone to come along and say, "Hey, let's get rid of the abrasion, let's get rid of the indirect process. Let's make the marks with ink right on paper instead."

But certainly Dataproducts was too slow to respond. I think we should have done it in partnership with the Japanese. If we'd wanted to go the laser route, we didn't have the xerographic technology. In America there was no mass producer of small laser engines. There were great producers of big engines, such as Xerox and Kodak. But not small engines. The small engine was needed for that printer. But even before lasers, in the time of dot matrix printers, we really never did compete with the Japanese in manufacturing. Dataproducts had some high quality, heavy-duty printers, and you see them even today in ticket-printing applications. American Airlines still has several thousand of those, at travel agents' desks around the country. Where you need a heavy-duty printer, those little light printers from Japan wouldn't stand up. But ours were priced in the two thousand dollar range as compared to their two hundred dollar printers. An order of magnitude different.

I'm not sure that there's something fundamental at work here. I think America can mass produce items in competition with the Japanese. I don't know that we can mass produce them in competition with the Philippines, though. It seems to me that one lives on an ever-shifting ground. The advantage the Japanese had they no longer have. Korea has it now, but won't have it very much longer. Singapore won't have it too much longer either. Perhaps Singapore is a little different. They've got an endless supply of cheap labor across the border. [laughter] As long as you have large sectors of the world that are relatively underdeveloped, there's always going to be a place with cheap labor.

I think we can compete with the Japanese today, and indeed isn't that happening? The Japanese are opening assembly plants here. They're now competing with plants in Japan. I don't have much trouble with the nationality of whoever owns the factory building if the money is being circulated here and the value is being added here. We get the jobs and the development of subcontractors and the improvement of the whole structure that occurs.

I think that there is a larger issue. The United States really faces some very difficult problems, as do the rest of the developed nations. We have enjoyed in this last century a much higher standard of living than the rest of the world. We feel we merited that because of the industrial revolution and its products and by-products. We have furnished goods and services that were desired by the rest of the world, for which they were willing to pay a premium. They have furnished raw material, for which we did not have to pay a premium. I think two things are going to happen: as the population increases and the competition for materials increases, their value will go up; and second, the costs of industrialization are equalizing.

The true cost of industrialization is just emerging as we begin to understand the environmental impact. We really ought to put that cost in the price of the product, but we don't know how to do that. It is not a lack of will so much as ignorance. When we sell a gallon of gasoline, we know that it's going to create pollution and that we're going to pay later to try to clean it up. We really ought to put those costs in the price.

What I think is happening in these last ten years, and at an accelerating rate, is that we are seeing a lowering of the standard of living in the developed countries due to these forces. It's not a temporary phenomenon due to an incorrect political policy or cyclical recession, but a real basic societal change. There is no reason why we should be permitted to maintain a superior standard of living if we can't produce desired goods and services for people. If they can do these things for themselves and are willing to live at a lower standard of living, our standard will inevitably have to come down. We won't deserve or get that premium. Don't despair, yet. I feel any individual will be able to stave this off for a while. It won't be really apparent for decades. Perhaps a century is a better time span to see what's happening. Some historian will be sitting here in the year 2200 and talking about what happened in the major shift of the year 2000.